Synthesis of Heterometallic Zinc-Gold and Lanthanide

Complexes and Reactivity Study of Pentaphosphaferrocene Towards Low-Valent

Main Group Species

Synthesis of Heterometallic Zinc-Gold and Lanthanide-Transition Metal Carbonyl Complexes and Reactivity Study of Pentaphosphaferrocene Towards Low-Valent Main Group Species

Zur Erlangung des akademischen Grades eines

DOKTORS DER NATURWISSENSCHAFTEN

(Dr. rer. nat.)

der KIT-Fakultät für Chemie und Biowissenschaften

des Karlsruher Instituts für Technologie (KIT)

vorgelegte

DISSERTATION

von

M.Sc. Ravi Yadav

aus

Haryana, India

KIT-Dekan: Prof. Dr. Reinhard Fischer

Referent: Prof. Dr. Peter W. Roesky

Korreferent: Prof. Dr. Annie K. Powell

Tag der mündlichen Prüfung: 17.10.2019

I hereby solemnly declare that I have authored the present work independently. I have not used sources other than those specified and cited in the bibliography. This thesis has not been submitted to any other university.

The presented work was carried out in the period from 01.07.2016 to 04.09.2019 at Institute of Inorganic Chemistry in Karlsruhe Institute of Technology (KIT) under the supervision of Prof. Dr. Peter W. Roesky. This work was funded by SFB1176.

Bibliografische Information der Deutschen Nationalbibliothek

Die Deutsche Nationalbibliothek verzeichnet diese Publikation in der Deutschen Nationalbibliografie; detaillierte bibliographische Daten sind im Internet über http://dnb.d-nb.de abrufbar.

1. Aufl. - Göttingen: Cuvillier, 2019

Zugl.: Karlsruhe (KIT), Univ., Diss., 2019

© CUVILLIER VERLAG, Göttingen 2019

Nonnenstieg 8, 37075 Göttingen

Telefon: 0551-54724-0

Telefax: 0551-54724-21

www.cuvillier.de

ISBN 978-3-7369-7123-3

eISBN 978-3-7369-6123-4

Table of Contents

1. Introduction

1.1 Metal complexes

Coordination compounds have been known since the 18th century.[1] However, their physicochemical properties were not well understood due to the absence of a proper theory. For example, the nature of ammonia ligand was not clear in complex CoCl$_3$·6NH$_3$. In 1893, Alfred Werner proposed the structure of CoCl$_3$·6NH$_3$ as an octahedral [Co(NH$_3$)$_6$]Cl$_3$ complex (A, Figure 1.1).[2] This hypothesis was the inception of Werner's theory, which allowed a new way of interpreting the presence of coordinating ligands around transition metals,[3] and resulted in an exponential increase in the study of transition metals in various ligand environments.[4] Later on, other theories, for example, the crystal field theory, the molecular orbital theory, and the ligand field theory were proposed to elucidate certain characteristics of some metal complexes which could not be explained by Werner's theory.[5] Although organometallic (hybrid of organic and inorganic chemistry) complexes were known since 1827,[6] the boost in organometallic chemistry started around 1950 after the discovery of landmark complexes such as ferrocene[7] and Fischer-carbenes[8].

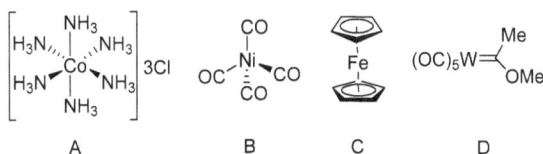

Figure 1.1: Selected examples of some milestone metal complexes; A: Werner's structure of CoCl$_3$·6NH$_3$; B: first binary metal-carbonyl complex; C: ferrocene; D: first Fischer-carbene complex.

1.2 Multinuclear metal complexes

Complexes containing more than one metal are referred to as di- or multinuclear complexes. A very broad library of multinuclear complexes has been accessed and consists of both homo- and heterometallic systems. Multinuclear complexes are interesting in several regards as they may, for example, promote cooperativity between different metals in one system, provide structural motifs known as metalloligands for the design of specific materials, and alter the physical

1

properties of the system.[9] Depending on the ways to arrange two or more metals in a complex, multinuclear complexes can be divided into two categories: i) complexes with a direct metal-metal bond, ii) two or more metals connected by bridging ligands.

1.2.1 Multinuclear complexes with direct metal-metal bond(s)

In 1938, the solid-state structure of [Fe$_2$(CO)$_9$] was determined by X-ray diffraction studies and, for the first time, the concept of metal-metal bond was proposed in a molecular complex.[10] The diamagnetic character of [Fe$_2$(CO)$_9$] can be explained by the pairing of two electrons on the Fe atoms, however, some theoretical arguments favour spin coupling via the carbonyl bridges without a direct Fe-Fe bond.[11] The concrete evidence of direct metal-metal bond came in 1957 via the solid-state structure determination of [Mn$_2$(CO)$_{10}$], the latter featuring an unsupported Mn-Mn single bond (Figure 1.2).[12]

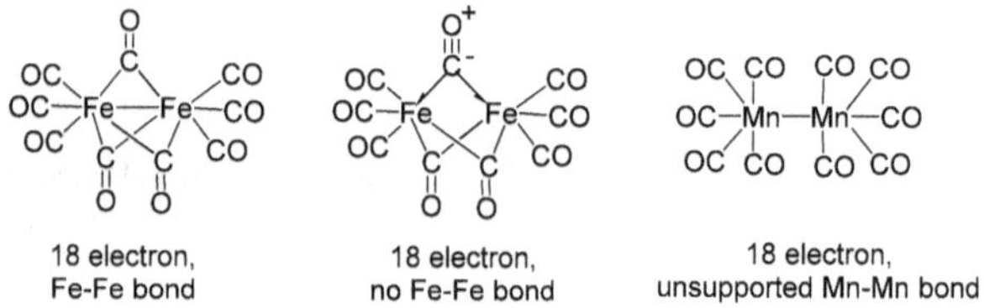

| 18 electron, Fe-Fe bond | 18 electron, no Fe-Fe bond | 18 electron, unsupported Mn-Mn bond |

Figure 1.2: Structural representation of Fe$_2$(CO)$_9$ and Mn$_2$(CO)$_{10}$.[11-12]

The breakthrough result in the field of metal-metal bonds was published in the early 1960s by Cotton, where the first quadruple bond was structurally characterized in the [Re$_2$Cl$_8$]$^{2-}$ ion.[13] The four unpaired electrons in the 5d orbitals of Re^{3+} are paired together to form a quadruple bond. After this work, a spiked interest in this field eventually resulted in metal-metal bonds with bond orders ranging from single to quintuple.[10] The bonding situation in such complexes was investigated to get an in depth knowledge of metal-metal bonds. An overview of the bonding situation in quadruple bonds (vide infra) has been discussed by Cotton and others (Figure 1.3).[10] Complexes with metal-metal bonds of lower bond order can also be described in a similar manner. To form a quadruple bond, each metal must have at least four unpaired electrons in the

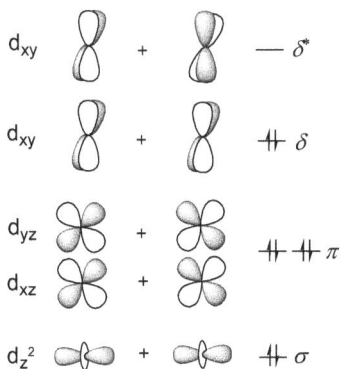

Figure 1.3: Overlap of d-orbitals involved in quadruple bond formation in [Re$_2$Cl$_8$]$^{2-}$ ion and the resulting energy levels.[10]

outermost d-orbital. For example, in the [Re$_2$Cl$_8$]$^{2-}$ ion, each Re atom is in the +3 oxidation state and possesses four unpaired electrons that can lead to bond formation with the other Re atom. The positive axial overlap between two d$_{z^2}$ orbitals results in the formation of a sigma bond (σ) and the corresponding negative overlap leads to an antibonding orbital (σ*). The lateral overlap between two d$_{xz}$ orbitals on the one hand and two d$_{yz}$ orbitals on the other hand, resulting in d$_{xz}$ + d$_{xz}$ and d$_{yz}$ + d$_{yz}$ combinations, gives rise to degenerate orthogonal π-orbitals. The corresponding negative overlap results in the corresponding antibonding π*-orbitals. Finally, the lateral overlap between d$_{xy}$ orbitals results in the formation of δ and δ* orbitals. The relative

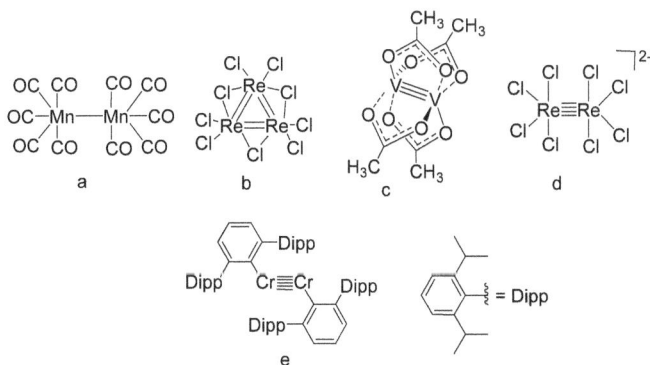

Figure 1.4: Selected examples of metal complexes with a metal-metal bond order from one to five.[10,14]

3

energy of the molecular orbitals can be determined using the Hückel concept *i.e.* the more the overlap, the lower the energy of the bonding molecular orbital. The overlap strength is increasing in the order $\delta < \pi < \sigma$, resulting in the following order in the energy of the molecular orbitals: $\sigma < \pi < \delta < \delta^* < \pi^* < \sigma^*$. The bond order corresponds to half of the difference between the numbers of bonding and antibonding electrons. In case of the $[Re_2Cl_8]^{2-}$ ion, eight electrons occupy bonding orbitals (excluding the Cl ions) and no electron is present in antibonding orbitals, resulting in a quadruple bond between the two Re atoms.[10] In 2006, Power and co-workers raised the bar for the maximum number of bonds possible between two elements by making the first quintuple bond in a chromium complex (e, Figure 1.4).[14]

1.2.2 Multinuclear complexes *via* bridging ligands

Using a bridging ligand is a common approach to make multinuclear complexes, the key aspect of this approach is ligand design rather than considerations of possible metal-orbital overlaps. In addition, this approach allows an easy inclusion of different types of metals in one complex. Multinuclear complexes can be classified into two groups a) homometallic complexes: two or more identical metals in one complex b) heterometallic complexes: two or more different metals in one complex.[15] Homometallic complexes can be conveniently synthesized by using multidentate ligands having similar coordination sites. On the other hand, the synthesis of heterometallic complexes requires specifically designed ligands having donor sites for binding to specific metals. Heterometallic complexes present the additional possibility of a synergistic reactivity, which is different than that observed for complexes containing only one type of metal.[16] Several types of heterometallic complexes have been reported in the literature and a very broad spectrum of ligands is available to synthesize them. In this section, only bidentate ligands featuring both a soft and a hard donor centre, in respect with the "hard and soft acids and bases" (HSAB) principle, are discussed. The most frequently used hard donor functional groups include carboxylates, alkyl or aryloxides, and amides, while typical soft donor groups involve thiols, phosphines, and carbenes. In the periodic table, early transition metals are classified as relatively hard acids, in comparison to the softer nature of late transition metals. Ligands with hard and soft donor groups within the same framework have been successfully employed to access early-late heterobimetallic (ELHB) complexes.

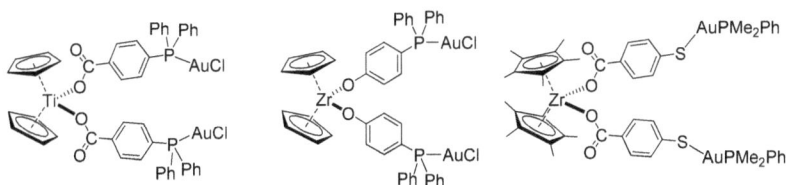

Figure 1.5: Selected examples of group-4 and [Au] heterobimetallic complexes.[17]

Apart from being fundamentally interesting, ELHB complexes have shown potential applications in catalysis, small molecule activation and medicinal chemistry.[18] Notably, P-N based ligands have been successfully employed for the synthesis of several ELHB complexes,[18-19] which have also been further used in various small-molecule activation. Especially, Ti/Zr-Co complexes supported by phosphinoimide ligands have been used to activate H_2,[20] CO_2,[21] organoazides,[22] and diazo reagents.[23] Very recently, a Zr-Co complex was used to activate dry oxygen and pyridine-N-oxide, and the obtained complexes were further tested for the ring opening of thf in the presence of Lewis acids. The use of P-N ligands to access heterometallic complexes is not only limited to transition metals as these ligands have also been used to synthesize lanthanide-transition metal complexes.[24] On the basis of the HSAB principle, hydroxy or carboxy group substituted phosphine ligands, that can be conveniently synthesized, are also promising supporting ligands for the selective preparation of heterometallic complexes. Corresponding heterometallic complexes of group 4 metals and gold have been synthesized and have shown interesting properties such as photoluminescence and anti-cancer properties (Figure 1.5).[17]

1.3 Divalent lanthanides

1.3.1 General

Lanthanides (Ln) belong to f-block elements and their most common oxidation states are 0 and +3, although the oxidation states +2 and +4 are also known depending on the Ln element.[25] The chemistry of trivalent lanthanides is very well explored and the corresponding complexes have found applications in various fields ranging from material science to bioinorganics.[25] In comparison, divalent lanthanides (Ln^{II}) are far less studied because the +2 oxidation state of lanthanides is not naturally occurring. Divalent lanthanide complexes are classified into two

5

classes *i.e.* classical and non-classical. Among the lanthanides, Eu, Yb, and Sm are the most stable in divalent oxidation state and are categorized as classical divalent lanthanides.[26] The remaining lanthanides are classified as non-classical divalent lanthanides Among the non-classical Ln^{II}, halides of only Tm, Dy, and Nd have been isolated in the solid state.[27] However, stable organometallic complexes of all the non-classical divalent lanthanides have also been successfully synthesized by employing bulky cyclopentadienyl ligands.[28] Apart from numerous applications of Ln^{II}, the use of $[Sm^{II}X_2]$ (X = Cl, Br, I) as a single-electron reductant in several organic reactions is a standout application.[29]

In 1964, Fischer reported the synthesis of the first organometallic complexes of divalent Eu and Yb by reaction of elemental Eu or Yb and cyclopentadiene in liquid ammonia.[30] Five years later, the first organometallic complex of divalent samarium, $[Cp_2Sm^{II}]$ (Cp = C_5H_5), was also obtained by reduction of $[Cp_3Sm^{III}]$ with potassium-naphthalenide in thf.[31] However, at that time, the interest in this field kept rather limited because of the insolubility of $[Cp_2Ln^{II}]$ (Ln = Sm, Eu, Yb) in common organic solvents. In 1981, Evans *et al.* synthesized the hydrocarbon-soluble divalent lanthanide complex $[Cp^*_2Sm^{II}(thf)_2]$ (Cp* = C_5Me_5) by vaporization of elemental Sm into a hexane solution of pentamethylcyclopentadiene.[32] Later on, a simple synthesis of $[Cp^*_2Sm^{II}(thf)_2]$ in solution was reported by reaction of $[Cp^*K]$ with $[SmI_2]$ in a 2:1 molar ratio in thf.[33] As of now, a substantial amount of research has been done on divalent lanthanocenes due to their easy accessibility, and especially on samarocene due to its highly reductive nature. It should be noted that $[Cp^*_2Sm^{II}(thf)_2]$ is a stronger reducing agent than $[Sm^{II}I_2]$ because Cp* ligands are more electron donating compared to iodides.[34] Sublimation of $[Cp^*_2Sm^{II}(thf)_2]$ at elevated temperature leads to the loss of the two thf molecules, resulting in the unsolvated samarocene, $[Cp^*_2Sm^{II}]$.[35] The most striking reactivity of $[Cp^*_2Sm^{II}]$ is the activation of dinitrogen at room temperature. The reaction of two molecules $[Cp^*_2Sm^{II}]$ with one molecule of N_2 occurs *via* single-electron reduction steps, resulting in the oxidation of Sm from +2 to +3, and the formation of a di-reduced N_2^{2-} dianion which is trapped in between two $[Cp^*_2Sm^{III}]^+$ moieties.[36] Roesky and co-workers have synthesized the first molecular f-block polyphosphide by activation of white phosphorous with $[Cp^*_2Sm^{II}]$.[37] However, no reaction was observed with the heavier pnictogens

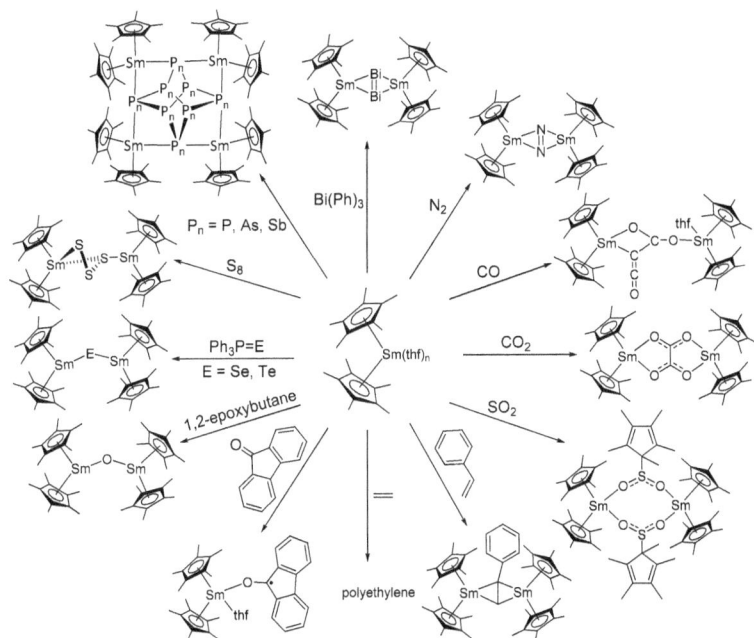

Figure 1.6: Reactivity of $[Cp*_2Sm^{II}(thf)_n]$ towards different organic and inorganic species (or elements).

such as elemental arsenic and antimony. Synthesis of the largest polyarsenides and polystibides of samarium was nevertheless possible by reaction of $[Cp*_2Sm^{II}]$ with highly reactive nanoscale arsenic[38] or nanoscale antimony,[39] respectively. The strong reductive ability of samarocene has also allowed the activation of small molecules such as CO_2,[40] CO,[41] SO_2,[42] and N_2.[41] Miscellaneous reduction reactions of main group species and elements by samarocene resulted in the isolation of several elusive anions such as $(P_n)_8^{4-}$ $(P_n = P, As, Sb)$,[37-39] Bi_2^{2-},[43] $[OC_3O_2]^{2-}$,[41] S_3^{2-},[44] and N_2^{2-}.[36] The reduction chemistry of samarocene has also been extended to organic molecules containing unsaturated bonds such as styrene and ethene (Figure 1.6).[45]

The chemistry of divalent lanthanides is dominated by the use of cyclopentadiene as a supporting ligand. The metal-ligand interaction involving f-block metals is more ionic in nature, compared to that with transition metals. Indeed, Wilkinson showed that the reaction of $[Cp_3Sm^{III}]$ with $FeCl_2$ resulted in the formation of ferrocene and samarium(III) chloride.[25] Due to this fact, it was long

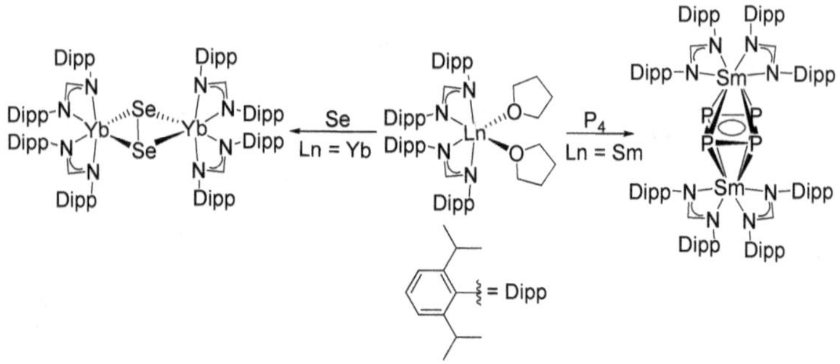

Figure 1.7: Reactivity of [(DippForm)$_2$Ln^{II}(thf)$_2$] towards white phosphorous and elemental selenium.[46]

considered that the nature of the ligands does not play a major role in the reactivity of lanthanide complexes. However, recent studies showed that the reactivity of divalent lanthanides may be altered when using different ligand environments (*vide infra*).[47] Amidinate ligands, which correspond to N-donor ligands, are interesting candidates due to their versatility in transition metal complexes and main group chemistry. In this context, Deacon and Junk have reported the synthesis of the divalent lanthanide complexes [(DippForm)$_2$Ln^{II}(thf)$_2$] (Ln = Sm,[48] Yb[49]) (DippForm = *N,N'*-bis(2,6-diisopropylphenyl)formamidinate), stabilized by bulky amidinate ligands. The reactivity of [(DippForm)$_2$Ln^{II}(thf)$_2$] contrasts with that of [Cp*$_2$Sm^{II}(thf)$_n$] towards pnictogens and chalcogens. The reaction of [Cp*$_2$Sm^{II}(thf)$_n$] with white phosphorous resulted in the cage like polyphosphide cluster [{Cp*$_2$Sm^{III}}$_4$(P)$_8$] (Figure 1.6).[37] Instead, [(DippForm)$_2$Sm^{II}(thf)$_2$] reacted differently in the same reaction, resulting in the formation of a planar P$_4^{2-}$ ring bridging two [(DippForm)$_2$Sm^{III}]$^+$ moieties (Figure 1.7).[46a] In addition, a different reactivity between the two Sm^{II} reagents was observed with elemental sulfur. The reaction of samarocene with sulfur or {Ph$_3$P=Se} resulted in the formation of [{Cp*$_2$Sm^{III}}$_2$S$_3$] with a bridging (S$_3^{2-}$) or [{Cp*$_2$Sm^{III}}$_2$Se] with a bridging (S$_3^{2-}$) or (Se^{2-}) unit, respectively (Figure 1.6).[44] Deacon, Junk, and co-workers have isolated (Se$_2^{2-}$) as bridging anion between two [(DippForm)Yb^{III}]$^{2+}$ moieties by reacting [(DippForm)$_2$Yb^{II}(thf)$_2$] with elemental selenium (Figure 1.7).[46b] In the case of [(DippForm)$_2$Ln^{II}(thf)$_2$] (Ln = Sm, Yb), a polysulfide cluster, [{(DippForm)Ln^{III}(S)$_4$}$_3$], was formed, resulting from the loss of one DippForm ligand.[50] The different reactivity of samarocene compared to [(DippForm)$_2$Ln^{II}(thf)$_2$] (Ln = Sm, Yb) may be explained by different steric and electronic environments around the lanthanide metal centre, resulting in a different stabilization of possible intermediates.

1.3.2 Lanthanide-Transition metal carbonyl complexes

F-block elements can be characterized by their strong Lewis acidity, as well as specific photoluminescence and magnetism properties.[25,51] Transition metal carbonyl complexes, on the other hand, are well known for their catalytic properties.[6] Therefore, heterometallic complexes of the type Ln-TM (TM = transition metal) carbonyl are interesting from both fundamental and application points of view.[52] Understanding the metal-metal interaction between d- and f-block elements is especially desirable. Ln-TM carbonyl complexes can be divided into three major categories: (i) isocarbonyl linkage between both metals, (ii) direct Ln-TM bond, and (iii) solvent-separated ion pairs.[52] Owing to the oxophilic nature of lanthanides and the stabilization of the anionic charge on the TM due to π-acceptor nature of the CO ligands, isocarbonyl bridged Ln-TM complexes are predominant over the two other systems.[53] Depending on the oxidation state of the lanthanide metal centre, two types of Ln-TM carbonyl complexes have been synthesized: i) Ln^{III}-TM carbonyl complexes and ii) Ln^{II}-TM carbonyl complexes. Ln^{III}-TM carbonyl complexes have been synthesized either by a salt metathesis reactions between trivalent lanthanide complexes and alkali-salts of TM carbonyls[54] or through redox reactions between divalent lanthanides and TM carbonyls.[55] In contrast, Ln^{II}-TM carbonyl complexes have been accessed by reaction of elemental lanthanides (Sm, Yb, Eu) with TM carbonyls in the presence of mercury.[52-53,56]

Anderson and co-workers reported the first structurally characterized Ln^{III}-TM carbonyl complex, [{$Cp*_2Yb^{III}$}$_2(\mu_2$-η^2-CO)$_4${(CO)$_7$Fe$_3$}], obtained by reduction of [Fe$_2$(CO)$_9$] with [$Cp*_2Yb^{II}$(OEt$_2$)].[55a] This strategy to reduce TM carbonyl complexes with highly reductive divalent lanthanide complexes was further applied to other TM metal complexes such as [Mn$_2$(CO)$_{10}$], [Co$_2$(CO)$_8$], and [{$Cp*$Fe(CO)$_2$}$_2$] (Figure 1.8).[55a-d,57] In such reactions, the divalent lanthanide complexes are oxidized from the +2 to +3 oxidation state *via* a single-electron transfer to the TM carbonyl moiety. As a result, formation of the [TM(CO)$_x$]$^-$ anion occurs, along with TM-TM bond cleavage. However, in some cases, instead of TM-TM bond cleavage, formation of new TM-TM bonds or new anionic TM carbonyl clusters is observed. For example, Deacon and co-workers isolated the

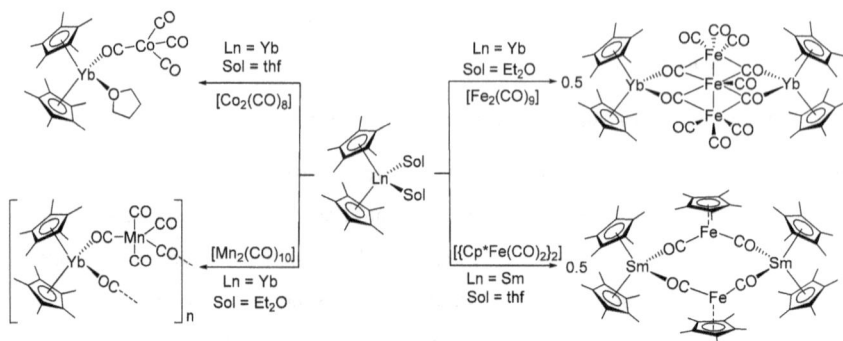

Figure 1.8: Ln^{III}-TM carbonyl complexes *via* reduction of TM carbonyls by divalent lanthanocenes.[55a-d]

elusive [W$_2$(CO)$_{10}$]$^{2-}$ anion, featuring an unsupported W-W bond in a mixed-valent Sm$^{II/III}$ calix[4]pyrrolide sandwich by reduction of [W(CO)$_6$] with a divalent samarium *meso*-octaethylcalix[4]pyrrolide.[58] The reduction of TM carbonyls by using different ligands around Ln^{II} have also been investigated. The reaction of [(Tp^{Me,Me})$_2$Sm^{II}] (Tp^{Me,Me} = hydro-tris-(3,5-dimethyl)pyrazolylborate) with [Re$_2$(CO)$_{10}$] resulted in the formation of [(Tp^{Me,Me})$_2$Sm^{III}][HRe$_4$(CO)$_{17}$] featuring a novel rhenium carbonyl cluster, [HRe$_4$(CO)$_{17}$]$^-$ (Figure 1.9).[59] It should be noted that divalent lanthanides have been also used in the reduction of TM carbonyl sulfides or polyphosphides. For example, the reduction of [Fe$_2$(μ-S$_2$)(CO)$_6$] by [Cp*$_2$Ln^{II}(thf)$_2$] (Ln = Sm, Yb) resulted in a wheel-shaped Ln-Fe sulfide cluster, [{Cp*$_2$Ln^{III}}$_2${(μ-S)$_6$(CO)$_{12}$Fe$_6$}].[60]

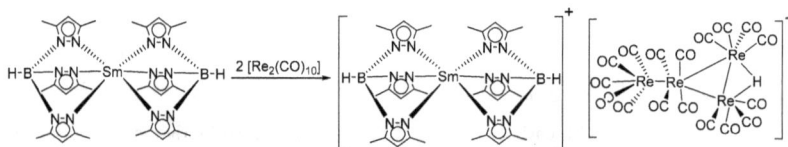

Figure 1.9: Synthesis of rhenium-carbonyl cluster by reduction of [Re$_2$(CO)$_{10}$] with [(Tp^{Me,Me})$_2$Sm^{II}].[59]

The first Ln^{II}-TM carbonyl complex, [(NH$_3$)$_n$Yb^{II}((μ-CO)$_4$Fe)], has been obtained by Shore and co-workers by reduction of [Fe$_3$(CO)$_{12}$] with elemental Yb in liquid ammonia. A ladder-type polymeric complex, [{((CH$_3$CN)$_3$Yb^{II}((μ-CO)$_4$Fe)}$_2$·(CH$_3$CN)]$_n$, with a direct Yb^{II}-Fe bond was obtained after crystallization from acetonitrile.[61] This Yb^{II}-Fe carbonyl complex can be considered as a lanthanide analogue of Colman's reagent, [Na$_2${Fe(CO)$_4$}]. Several other Ln^{II}-TM carbonyl complexes have been prepared by either reduction of TM carbonyl complexes with Ln/Hg amalgam or by redox-transmetallation between Hg salts of TM carbonyl complexes and elemental Ln^0.[49,52-53,56,62]

1.4 Pentaphosphaferrocene

Ferrocene is arguably one of the most important discoveries in organometallic chemistry of last century.[7] This sandwich compound has attracted a lot of attention due to its exciting physical and chemical properties, and has found applications in numerous fields such as synthetic chemistry, polymer chemistry, biomedical and material science.[63] The specific properties of metallocenes can be fine-tuned by using derivatives of the Cp$^-$ ligand, which has prompted the search for analogues of the Cp$^-$ ring that can be used to make new metallocenes. Tilly, Sekiguchi, Driess, and Saito have independently reported the synthesis of heavier group-14 analogues of Cp$^-$.[64] Derivatives of Cp$^-$ are not limited to group-14 elements, and other 5-membered rings containing heteroatoms such as [CB$_2$N$_2$]$^-$,[65] [C$_3$B$_2$]$^-$,[66] [PC$_4$]$^-$,[67] [SbC$_4$]$^-$,[68] and [BiC$_4$]$^-$,[69] have been reported. Taking into account the isolobal analogy between Cp$^-$ and cyclo-P$_5^-$, the latter was proposed to be a suitable ligand for the preparation of metallocenes.[70] In a seminal report, Scherer and co-workers have synthesized [Cp*Fe(η^5-P$_5$)], the first double-decker metallocene bearing the cyclo-P$_5^-$ ring as a ligand, by co-thermolysis of [{Cp*Fe(CO)$_2$}$_2$] with white phosphorous (Figure 1.10).[71]

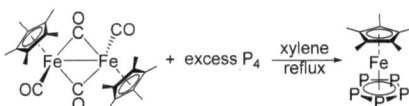

Figure 1.10: Synthesis of [Cp*Fe(η^5-P$_5$)] by co-thermolysis of [{Cp*Fe(CO)$_2$}$_2$] and P$_4$.[71]

1.4.1 Coordination chemistry of pentaphosphaferrocene

[Cp*Fe(η^5-P$_5$)] has revealed to be a very useful tool in coordination chemistry due to the presence of five free lone-pairs on the cyclo-P$_5$ ring.[72] Using [Cp*Fe(η^5-P$_5$)] as a metalloligand, Scheer et al. have carried out pioneering work in the field of inorganic supramolecular chemistry. For example, the reaction of [Cp*Fe(η^5-P$_5$)] with Cu^{I}X (X = Br, I) resulted in the 2D coordination

11

polymers [{Cp*Fe(η^5-η^1-η^1-η^1-P$_5$)}(Cu^{I}X)]$_n$ (X = Br, I).[73] However, reaction of [Cp*Fe(η^5-P$_5$)] with Cu^{I}Cl led to the formation of two different products, [{Cp*Fe(η^5-η^1-η^1-P$_5$)}(Cu^{I}Cl)]$_n$, corresponding to a 1D polymer, and [{Cp*Fe(η^5-η^1-η^1-η^1-η^1-η^1-P$_5$)}$_{12}$(Cu^{I}Cl)$_{10}${Cu^I(CH$_3$CN)$_2$}$_5$], a fullerene-type supramolecule. The structural motif of [{Cp*Fe(η^5-η^1-η^1-η^1-η^1-η^1-P$_5$)}$_{12}$(Cu^{I}Cl)$_{10}${Cu^I(CH$_3$CN)$_2$}$_5$] is reminiscent of the C$_{60}$ arrangement: the *cyclo*-P$_5$ ring is analogous to the C$_5$ moiety and the P$_4$Cu$_2$ six-membered ring, obtained by coordination of [Cp*Fe(η^5-P$_5$)] to Cu^{I}Cl units, is analogous to the C$_6$ moiety in C$_{60}$.[74] The outer diameter of the resulting inorganic supramolecule is 2.13 nm, which is approximately three times the diameter of C$_{60}$. Furthermore, the inner diameter of the inorganic supramolecule, 1.25 nm, is also larger than that of C$_{60}$.[74-75] These large inorganic fullerene-type molecules were used to encapsulate several host molecules such as C$_{60}$,[76] [Cp*Fe(η^5-P$_5$)],[74] P$_4$,[77] As$_4$,[77] *o*-carborane,[78] *etc.* The coordination chemistry of [Cp*Fe(η^5-P$_5$)] towards silver salts has also been studied. Although reaction of [Cp*Fe(η^5-P$_5$)] with silver salts featuring classical weakly-coordinating anions (PF$_6^-$, BF$_4^-$, or CF$_3$SO$_3^-$) resulted in insoluble precipitates, a 1D coordination polymer, [{Cp*Fe(η^5-η^1-η^1-η^1-P$_5$)}$_2${AgAl{OC(CF$_3$)$_3$}$_4$}]$_n$, was isolated in the solid state when using [Al{OC(CF$_3$)$_3$}$_4$]$^-$ as counteranion.[79] In the solid-state structure, the silver cations are coordinated by the *cyclo*-P$_5$ ring of [Cp*Fe(η^5-P$_5$)] in both side-on and end-on coordination modes (Figure 1.11).

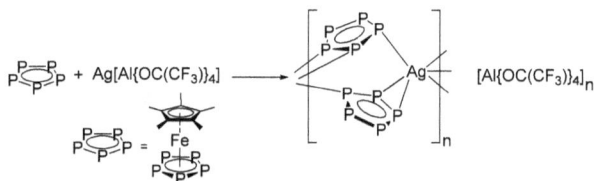

Figure 1.11: Formation of a 1D coordination polymer of Ag$^+$ salt and [Cp*Fe(η^5-P$_5$)].[79]

1.4.2 Reactivity of the *cyclo*-P$_5$ ring of pentaphosphaferrocene

Winter and Geiger studied the redox properties of [Cp*Fe(η^5-P$_5$)] by cyclic voltammetry (CV), showing irreversible 1e$^-$ redox events upon reduction or oxidation.[80] The irreversible nature of the redox processes can be explained by the generation of either a 19e$^-$ or a 17e$^-$ species, which then dimerize. The possibility to observe the formation of a 2e$^-$ reduction species was further investigated by Scheer and co-workers by CV experiments under various conditions. Unfortunately, the corresponding di-reduced complex could not be detected.[81] Roesky and co-workers succeeded in the synthesis of the first example of a formally di-reduced [Cp*Fe(η^5-P$_5$)] species by reaction of [(DIP$_2$pyr)Sm^{II}I(thf)$_3$] (DIP$_2$pyr = 2,5-bis{N-(2,6-diisopropylphenyl)iminomethyl}pyrrolyl) with [Cp*Fe(η^5-P$_5$)] in the presence of potassium metal (c, Figure 1.12).[82] The addition of two electrons to the [Cp*Fe(η^5-P$_5$)] moiety disrupts the 6π-electron aromaticity of the planar *cyclo*-P$_5$ ring. As a result, an envelope-shaped *cyclo*-P$_5$ ring, η^4-

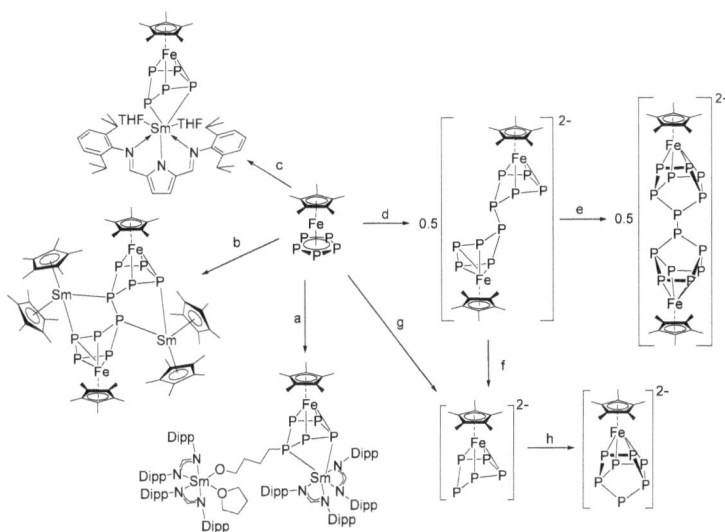

Figure 1.12: Reactivity of [Cp*Fe(η^5-P$_5$)] towards reducing agents: a) [(DippForm)$_2$Sm^{II}(thf)$_2$] in toluene; b) [Cp*$_2$Sm^{II}(thf)$_2$] in toluene; c) [(DIP$_2$pyr)Sm^{II}I(thf)$_3$] and K in thf; d) KH in thf; e) P$_4$ in dme; f) K in dme; g) excess K in dme; h) P$_4$ in dme.

coordinated to the [Cp*Fe]$^+$ moiety, is formed. Besides, the mono-reduced analogue, [(Cp*$_2$Sm^{III})$_2$(P$_{10}$)(Cp*Fe)$_2$] (b, Figure 1.12), could be synthesized by reduction of [Cp*Fe(η^5-P$_5$)] with [Cp*$_2$Sm^{II}(thf)$_2$], followed by a reductive P-P bond formation, as predicted by the early CV studies.[83] Interestingly, the reduction of [Cp*Fe(η^5-P$_5$)] with the more sterically encumbered [(DippForm)$_2$Sm^{II}(thf)$_2$] led to another reaction pathway. In this case, the trinuclear complex [(DippForm)$_2$Sm^{III}Cp*Fe(η^4-P$_5$){(CH$_2$)$_4$O}(DippForm)$_2$Sm^{III}(thf)] was formed through a ring opening of one thf molecule, resulting in a butanoate moiety bridging the

[(DippForm)$_2$Sm^{III}Cp*Fe(η^4-P$_5$)] and [(DippForm)$_2$Sm^{III}(thf)] moieties (a, Figure 1.12).[84] Scheer *et al.* have also engaged in a comprehensive study of the reduction and oxidation of [Cp*Fe(η^5-P$_5$)] (using KH or K as reducing agents, and a thianthrenium salt as oxidizing agent) and the corresponding products were further used to activate white phosphorous (d-h, Figure 1.11).[81] DFT calculations performed on [Cp*Fe(η^5-P$_5$)] showed a significant contribution of the P atoms of the *cyclo*-P$_5$ ring in the HOMOs and LUMOs, the latter also featuring metal contributions.[85] Such theoretical results are supported by the involvement of the *cyclo*-P$_5$ ring during redox event.

The reactivity of [Cp*Fe(η^5-P$_5$)] has also been studied using main-group nucleophiles. The reaction of [Cp*Fe(η^5-P$_5$)] with LiCH$_2$SiMe$_3$, LiNMe$_2$, NaNH$_2$, and LiPH$_2$ resulted in nucleophilic addition reactions on the *cyclo*-P$_5$ ring (Figure 1.13).[86] Reaction with the [CH$_2$SiMe$_3$]$^-$ and [NMe$_2$]$^-$ nucleophiles results in the formation of an envelope-shaped *cyclo*-P$_5$ ring accompanied by P-C or P-N bond formation. In the case of NaNH$_2$ and LiPH$_2$, bis- and tris-anionic derivatives were obtained by aggregation of the corresponding alkali metal salts (Figure 1.13). The resulting derivatives of pentaphosphaferrocene were further used in the synthesis of neutral triple deckers *via* salt-metathesis reactions between [Cp*Fe(η^4-P$_5$R)(Li)] (R = CH$_2$SiMe$_3$, NMe$_2$) and [Cp$'''$MX]$_2$ (Cp$'''$ = 1,2,4-tris(tertiary-butyl)cyclopentadienyl; M = Cr, Fe, Co, Ni; X = Cl, Br).[87]

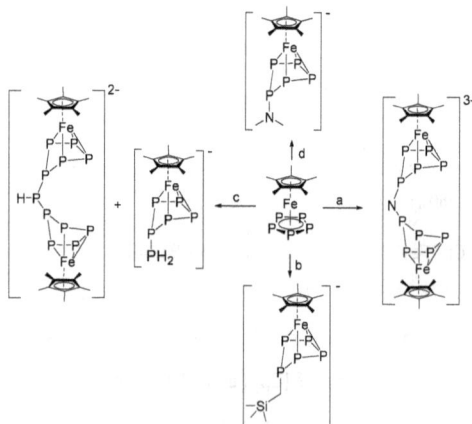

Figure 1.13: Reactivity of [Cp*Fe(η^5-P$_5$)] towards main group nucleophiles: a) NaNH$_2$ in dme; b) LiCH$_2$SiMe$_3$ in Et$_2$O; c) LiPH$_2$ in thf; d) LiNMe$_2$ in thf.[86]

1.5 Low-valent main group compounds

1.5.1 *N*-Heterocyclic carbenes

Carbenes are defined as divalent carbon species possessing two bonding valence electrons and two non-bonding valence electrons. In free carbenes, two non-bonding electrons occupy the non-bonding orbitals (NBOs) p_x and p_y, giving rises to i) a diamagnetic singlet carbene, when the electrons are paired up in a single orbital, leaving one orbital empty and generating $\sigma^2 p_\pi^0$ frontier molecular orbitals and ii) a paramagnetic triplet carbene, when two electrons are unpaired and occupy both NBOs, generating frontier molecular orbitals of the type $\sigma^1 p_\pi^1$ (Figure 1.14).[88]

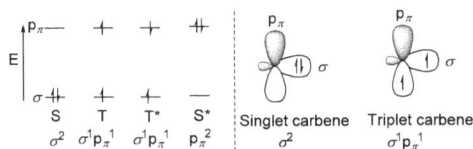

Figure 1.14: Representation of frontier molecular orbitals of carbenes.

In 1968, Wanzlick reported the synthesis of a mercury carbene complex without isolation of the elusive free carbene intermediate.[89] Almost 20 years later, Bertrand isolated the first free carbene by taking advantage of the stabilization effect of phosphorous and silicon substituents.[90] Extensive research in the chemistry of NHCs started after the isolation of the first stable "bottle-able" NHC, 1,3-bis(adamantyl)imidazol-2-ylidene, by Arduengo in 1991.[91] The presence of adjacent nitrogen atoms next to the carbenic centre results in a stabilization of the σ^2 orbital by a negative inductive effect. In addition, donation to the empty p_π orbital by a positive mesomeric effect results in the destabilization of the latter orbital. As a result, the energy difference between the filled σ^2 (HOMO) and empty p_π (LUMO) orbitals is increased, which favours a singlet ground state.[92] The overall σ- and π-donor and π-acceptor properties of NHC depend on the nature of the atoms in the heterocyclic ring as well as on the substituents on the nitrogen atoms.[93] Intense research in the field of carbenes has made it possible to access a wide range of NHCs with different substituents on the heterocyclic carbene ring, resulting in a variation of the thermodynamic properties and kinetic protection.[94] The coordination chemistry of NHCs with s-,[95] p-,[95] d-,[96] and f-block[97] elements has been rigorously studied and has led to several applications from medicinal use to catalysis.[94,98] The scope of NHCs is not only limited to coordination chemistry as NHCs have also been used to activate small molecules.[95]

1.5.2 *N*-Heterocyclic silylenes

Silylenes, which are divalent silicon species, correspond to the heavier analogues of carbenes. In principle, silylenes can exist in the triplet or singlet state. However, due to the large energy difference between the 3s and 3p orbitals of silicon, silylenes usually possess a singlet ground state, albeit bis(tri-*tert*-butylsilyl)silylene is a remarkable example of a silylene with a triplet ground state.[99] Until 1986, silylenes were considered as transient species and known to decompose even at very low temperatures (-196 °C).[100] In 1986, Jutzi reported the first molecular divalent silicon compound stable at room temperature, [Cp*$_2$Si^{II}], by the reduction of [Cp*$_2$SiCl$_2$] with sodium naphthalenide in thf. In 1994, Denk and West reported a di-coordinate NHSi (*N*-heterocyclic silylene), corresponding to the silicon analogue of NHCs, by reducing the corresponding *N*-heterocyclic silicon dichloride compound with elemental potassium.[101] Since then, a large number of silylenes, including 4-, 5-, and 6-membered *cyclic*- as well as *acyclic*-compounds, have been reported. Similarly to NHCs, the singlet ground state of NHSis benefits from the electron donation from the adjacent nitrogen atom/s to the vacant p$_\pi$ orbital. The presence of a lone pair and an empty p$_\pi$ orbital in their singlet ground-state renders NHSis both Lewis acidic and basic in nature. This property of NHSi has been utilized in small molecule activation and NHSis have been used to activate diverse range of small molecules such as CO_2, NH_3, PH_3, AsH_3, acetylenes, E (E = S_8, Se, and Te), hexafluoro-benzene, organo-azides, just to cite a few.[102]

1.5.3 Mono-valent aluminium

Aluminium is the most abundant metal in the earth's crust and the corresponding minerals can be processed to generate the metal in bulk scale, which is the main reason for its low price even considering the tremendous uses in several industries.[103] Replacing less-abundant and expensive transition metal complexes from catalysis is a major topic in current science, and aluminium is a suitable candidate.[104] The quest to mimic the behaviour of transition metals has led to extensive investigations in the chemistry of low-valent main-group elements in the past few decades.[105] Recently, the chemistry of low-valent aluminium compounds has witnessed a renewed interest which can be attributed to the ability of such species to activate small molecules and organic substrates having single, double, and triple bonds. The mono-valent aluminium compounds are electronically analogous to carbenes.[106] Schnöckel and co-workers isolated the first example of a molecular Al^I complex, $[(Cp*Al^I)_4]$, which crystallized in a tetrameric form.[107] Initially, $[(Cp*Al^I)_4]$ was synthesized by the condensation of metastable $(Al^ICl)_8$[108] onto a solution of $[Cp*_2Mg]$. However, this procedure requires a special condensation apparatus.[109] Later on, H. W. Roesky and co-workers reported a simplified procedure to synthesize $[(Cp*Al^I)_4]$ which involves the reduction of $[Cp*AlCl_2]$ by elemental potassium in hot toluene.[110] The tetrameric $[(Cp*Al^I)_4]$ complex shows an equilibrium with the monomeric form $[Cp*Al^I]$ at elevated temperature,[109] the latter form being the actual carbene analogue. Taking advantage of this equilibrium phenomenon, Roesky and co-workers reported the synthesis of $[Cp*_2Ln^{II}$-$Al^ICp*]$ (Ln = Yb and Eu) by co-thermolysis of $[Cp*_2Ln^{II}]$ and $[(Cp*Al^I)_4]$ under reduced pressure. In the isolated complexes, an unprecedented donor-acceptor Al-Ln bond was observed.[111] Later on, Arnold *et al.* have used a similar strategy to make donor-acceptor Al-U bonds using $[(Cp*Al^I)_4]$ as a starting material.[112] Besides, Fischer and co-workers have studied the coordination behaviour of $[(Cp*Al^I)_4]$ towards transition metals.[113] It should be noted that the chemistry of $[(Cp*Al^I)_4]$ is not limited to its use as a coordinating ligand. $[(Cp*Al^I)_4]$ has also been used to activate small molecules and main groups elements. In this context, the reaction of $[(Cp*Al^I)_4]$ with elemental chalcogens resulted in the formation of Al-chalcogenide cages of the type $[(Cp*AlE)_4]$ (E = O, S, Se, and Te).[106,110] $[(Cp*Al^I)_4]$ readily reacts with white phosphorous at room temperature furnishing an unusual aluminium phosphide cage like molecule, $[(Cp*Al)_6\{(\mu-\eta^3-P)_4\}]$. Schnöckel and co-workers have also synthesized several low-valent aluminium based clusters upon changing the Cp* ligand system such as $[N\{Si(Me)_3\}_2]^-$[114] and $[P(^tBu)_2]^-$.[115],[116] However, isolation of first monomeric complex of monovalent aluminium was possible by using sterically bulky [Dipp-*BDI*] ligand (Dipp-*BDI* = $(2,6-^iPr_2C_6H_3NCMe)_2CH$). H. W. Roesky reported the monomeric [Dipp-*BDI*-Al^I] complex by reducing [Dipp-*BDI*-AlI_2] with elemental potassium at room temperature.[117] Owing to the monomeric form [Dipp-*BDI*-Al^I] have a lone pair on the Al atom and is more reactive than $[(Cp*Al^I)_4]$ where the electrons are shared by Al atoms to form a tetrahedral arrangement. The reactivity of [(Cp*Al^I)$_4$] and [Dipp-*BDI*-Al^I] is also different possible due the more steric bulk and presence of lone pair at room temperature in the latter.[118] For example, the reaction of [Dipp-*BDI*-Al^I] white phosphorous resulted in the formation of an Al-P heterocyclic six membered ring [(Dipp-*BDI*-Al^{III})$_2$(μ-P$_4$)] *via* cleavage of two opposite P-P bonds of the P$_4$ tetrahedron.[119] Very recently, Braunschweig *et al.* reported a cyclopentadiene based [Al^I] complex, [Cp'''Al^I], exibhithing monomeric form at room temperature.[120]

2. Aim of the Project

The objectives of this thesis are the synthesis of heterometallic complexes of high nuclearity using different methodologies and the study of the reactivity of pentaphosphaferrocene towards different low-valent main group reagents. The synthesis of high nuclearity heterometallic complexes will be carried out following two approaches: i) by employing bifunctional ligands that can selectively bind to Zn and Au centres ii) by redox reactions between divalent lanthanide complexes and TM carbonyls. In the first part, Zn-metalloligands will be synthesized using bifunctional carboxy-phosphine ligands. The corresponding Zn complexes, which feature free phosphine moieties, will be utilized in further coordination chemistry with gold(I) metal centres to obtain Zn-Au heterometallic complexes. The role of the ligand in Au^I complexes is vital to facilitate aurophilic interactions. Therefore, in this work, two slightly different carboxy-phosphine ligands will be used to study their ability to promote aurophilic interactions in the corresponding Zn-Au heterometallic complexes.

In a second part, the highly reductive nature of Ln^{II} complexes will provide the opportunity to obtain Ln^{III}-TM carbonyl complexes *via* a direct synthesis involving reduction of TM carbonyls by Ln^{II} complexes. As research in this area is mainly limited to the use of [Cp*$_2$Yb^{II}(OEt$_2$)$_2$] as a reaction partner with the TM carbonyls (Section 1.3.2), the reactivity of 3d and 5d TM carbonyls will be studied towards the sterically encumbered divalent lanthanide complexes [(DippForm)$_2$Ln^{II}(thf)$_2$] (Ln = Sm, Yb). Hence, the effect of the ligand environment around the Ln^{II} centre on the reactivity of the corresponding complexes towards TM carbonyls will be investigated.

In a third part, the reactivity of [Cp*Fe(η^5-P$_5$)] towards low-valent main group compounds will be presented. Pentaphosphaferrocene has recently attracted a large attention and has been used for the synthesis of several inorganic polymers and supramolecules. So far, the reactivity of [Cp*Fe(η^5-P$_5$)] has only been reported with single electron reductants (KH, K, and Ln^{II} complexes) and with main group nucleophiles (NaNH$_2$, LiPH$_2$ *etc.*) (Section 1.4.2). In this work, the reactivity of [Cp*Fe(η^5-P$_5$)] will be investigated towards low-valent main group compounds, with the aim to substitute and insert other atoms in the *cyclo*-P$_5$ ring of [Cp*Fe(η^5-P$_5$)]. The replacement of

19

one or more P atoms by other main group elements would disrupt the five-fold symmetry of the cyclo-P$_5$ ring, which could provide new heteroatomic polyphosphide rings and cages, and open new scopes in the coordination chemistry of phosphaferrocene.

3. Results and Discussion

3.1 Heterometallic complexes of Zn and Au

3.1.1 Introduction

The study of heterometallic complexes is a forefront area in present day inorganic chemistry. Most of the metals have some unique properties and transition metals have especially attracted attention due to their ability to act as catalysts in numerous chemical transformations. In order to develop catalysts with very specific sets of properties and perform kinetically unfavourable chemical reactions, an increased demand for tailor made complexes has arisen.[9] The utilization of heterometallic complexes is one of the solutions. The development of methodologies for the design of new combinations of heterometallic complexes is a major goal and the use of bifunctional ligands with different binding sites, for example based on the Pearson acid-base concept, is an interesting approach.[17c]

Through the incorporation of phosphine and carboxylate groups within the same ligand framework, the access to a wide range of heterometallic complexes has been reported.[17b,121] Recently, carboxy-functionalized phosphine ligands have been used to synthesize Ru-Zn[122] or Sn-Au[123] heterometallic complexes. To the best of our knowledge, heterometallic complexes containing both zinc-carboxylate and gold-phosphine moieties have not been reported to date. These two types of coordination compounds are well documented and phosphine adducts of Au^{I}Cl have especially shown interesting catalytic and photophysical properties.[124] In order to synthesize Zn-Au heterometallic complexes, commercially available bifunctional carboxy-phosphine ligands were selected (Chart 3.1.1). The effect of spacer between the carboxy and the phosphine has been studied by using either a rigid phenyl ring or a flexible ethyl group.

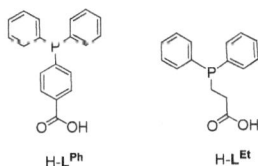

Chart 3.1.1 The carboxy-phosphine ligands used to synthesize Zn-Au heterometallic complexes.

21

3.1.2 Synthesis of a rigid zinc-metalloligand bearing terminal phosphines

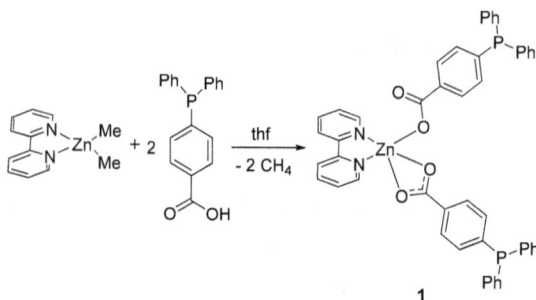

Scheme 3.1.1: Synthesis of complex 1.

For the synthesis of the zinc-based metalloligand **1**, an acid-base reaction was carried out between [(Bipy)ZnMe$_2$][125] (Bipy = 2,2'-bipyridine) and 4-(diphenylphosphino)benzoic acid (H-L^{Ph}) in a 1:2 molar ratio (Scheme 3.1.1). Reaction between both starting materials is indicated by a change in the colour of the reaction mixture, from light yellow to colourless, accompanied by the liberation of methane gas. After a short work up, complex **1**, [(Bipy)Zn(p-O$_2$C(C$_6$H$_4$)PPh$_2$)$_2$], was isolated as colourless crystals in 70% yield. The ^{1}H NMR spectrum of the product shows exclusive resonances in the aromatic region, ranging from δ 7.28 to 9.14 ppm, corresponding to aryl protons of the L^{Ph} and bipyridine ligands. In the ^{31}P{^{1}H} NMR spectrum, a singlet is detected at δ -5.22 ppm, which is in the usual range for uncoordinated triarylphosphines.[121c] The identity of complex **1** was further established by single-crystal X-ray diffraction studies. Complex **1** crystallizes in the triclinic space group $P\bar{1}$ with one molecule in the asymmetric unit cell (Figure 3.1.1). The solid-state structure shows that the Zn atom is pentacoordinated, surrounded by the two N donors of the chelating bipyridine and three O atoms of two carboxylate ligands. The Zn-N1 (2.0658(14) Å) and Zn-N2 (2.0872(12) Å) bond distances are in the typical range of bipyridine-zinc-carboxylate moieties (2.057(7)-2.100(8) Å).[126] Interestingly, the carboxylate groups of the two L^{Ph} ligands around the Zn centre exhibit different coordination modes *i.e.* mono- (κ^1) and bi-dentate (κ^2). The Zn-O3 (1.921(2) Å) bond length is slightly shorter than the Zn-O1 (1.9813(11) Å) analogue involving a mono-dentate (κ^1) coordination mode of the carboxylate.[127] In comparison, the Zn-O2 bond length (2.4781(13) Å) is much longer, which suggests a weaker coordination.

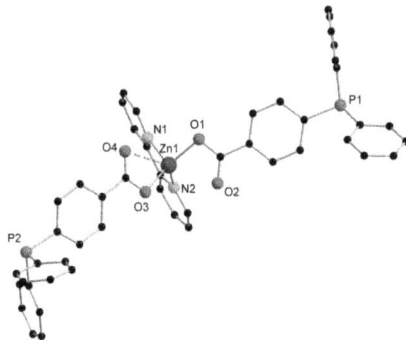

Figure 3.1.1: Molecular structure of complex **1**. Hydrogen atoms are omitted for clarity. Selected bond lengths (Å) and angles [°] : Zn-O1 1.9813(11), Zn-O2 2.4781(13), Zn-O3 1.921(2), Zn-N1 2.0658(14), Zn-N2 2.0872(12), O1-C30 1.275(2), O2-C30 1.240(2), O3-C11 1.287(2), O4-C11 1.230(2); O1-Zn-O2 57.73(4), O1-Zn-N1 101.77(6), O1-Zn-N2 121.08(5), O3-Zn-O1 127.40(5), O3-Zn-O2 101.72(5), O3-Zn-N1 116.12(6), O3-Zn-N2 101.76(6), N1-Zn-N2 78.89(6), O4-C11-O3 123.70(13), O2-C30-O1 121.66(13).

3.1.3 Synthesis of a nonanuclear Zn-Au heterometallic complex

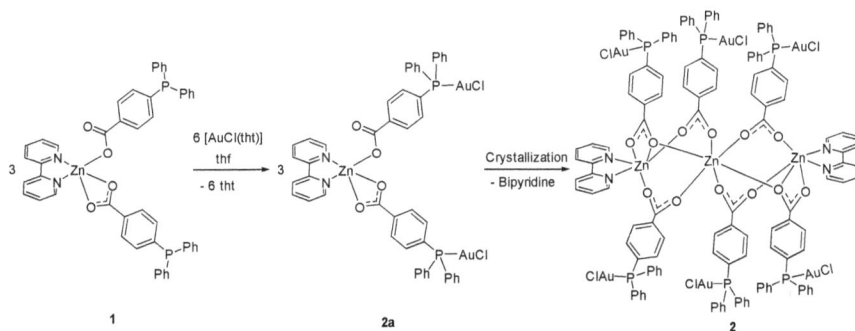

Scheme 3.1.2: Synthesis of complexes **2a** and **2**.

The zinc-based metalloligand **1** was reacted with [AuCl(tht)][128] (tht = tetrahydrothiophene) to obtain heterometallic complex. The reaction between **1** and [AuCl(tht)] in thf at room temperature led to the formation of [(Bipy)Zn(p-O$_2$C(C$_6$H$_4$)PPh$_2$-AuCl)$_2$] (**2a**), resulting from the coordination of the phosphine moiety to the gold centre (Scheme 3.1.2). This was confirmed by

23

analysis of the $^{31}P\{^1H\}$ NMR spectrum where a broad resonance is detected at δ 33.5 ppm, downfield shifted compared to complex **1**.[121c] In the ^1H NMR spectrum, the resonances for the aromatic protons are also slightly downfield shifted upon coordination of the phosphine groups to the [AuCl] fragment.[121c] The integration ratio of the different protons in the ^1H NMR spectrum as well as elemental analysis data are consistent with the proposed structure of **2a**. However, upon crystallization by slow evaporation of a solution of **2a** in a mixture of solvents (dichloromethane, acetone, and ethanol), a small amount of single crystals corresponding to the higher nuclearity complex {(Bipy)$_2$Zn$_3$(p-O$_2$C(C$_6$H$_4$)PPh$_2$-AuCl)$_6$} (**2**) were formed (Figure 3.1.2). Interestingly, complex **2** could not be detected in the ^1H NMR spectrum of the crude reaction mixture before the crystallization step. Complex **2** results from the assembly of three molecules of **2a** along with the de-coordination of one bipyridine ligand. It should be noted that the loss of coordinating ligands in zinc-carboxylate systems is a known phenomenon.[129] Complex **2** crystallizes in the triclinic space group $P\bar{1}$ with half a molecule in the asymmetric unit cell. Each Zn atom is in a distorted octahedral environment. Both Zn1 and Zn1' are coordinated by the two N atoms of a chelating bipyridine and four O atoms belonging to three different **L^{Ph}** carboxylate groups. In contrast, Zn2 is exclusively coordinated by O atoms, from six different carboxylate groups. The carboxylate moieties show two different coordination modes, κ^2 and κ^3. The Zn-N bond distances in **2** are longer than those in complex **1** (2.140 *vs* 2.076 Å, respectively), which can be attributed to the increase in the coordination number of the Zn atoms (6 *vs* 5, respectively) leading to a weaker coordination of the carboxylate group (Zn1-O bond distance in the range 2.021(8)-2.281(8) Å for complex **2**). The average P-Au bond length (2.231 Å) and the Au-Cl bond distances (ranging from 2.283(3) to 2.294(4) Å) are in agreement with the reported values for triarylphosphine-AuCl adducts.[130] At first glance, the reason for the formation of the nonanuclear assembly in **2** might be attributed to aurophilic interactions. However, intra- or inter-molecular Au-Au interactions (Au-Au distance: 2.70-3.50 Å)[130a,131] are not observed in the solid-state structure of complex **2**, probably because of steric crowding and rigidity of the triarylphosphine ligand. The possible reason for the formation of **2** from **2a** might be due to crystal packing effects. Unfortunately, during the crystallization process, crystals of **2** were formed along with a significant amount of amorphous material corresponding to **2a**, which hinders clean isolation and full characterization of **2**. All the attempts to selectively crystallize either **2a** or **2** by using different crystallization techniques only resulted in amorphous material containing **2a**.

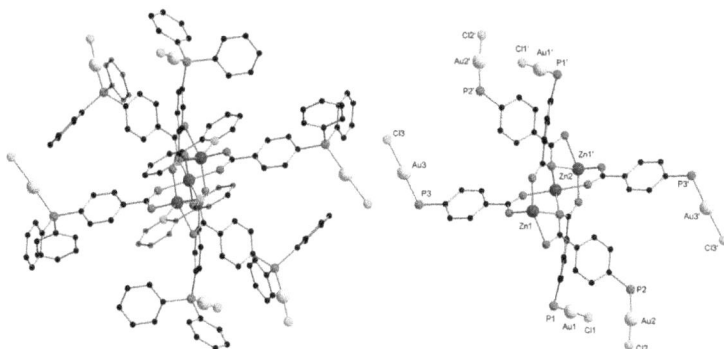

Figure 3.1.2: Molecular structure of complex **2** (left) and clear view of the structure after omitting the phenyl groups of the phosphines and the bipyridines (right). Hydrogen atoms are omitted for clarity. Selected bond lengths (Å) and angles [°]: Au1-Cl1 2.283(3), Au2-Cl2 2.291(7), Au3-Cl3 2.294(4), Au1-P1 2.221(3), Au2-P2 2.246(5), Au3-P3 2.226(4), Zn1-N1 2.119(9), Zn1-N2 2.162(9), Zn1-O1 2.281(8), Zn1-O2 2.145(8), Zn1-O3 2.021(8), Zn1-O6 2.029(8), Zn2-O2 2.183(8), Zn2-O4 2.071(7), Zn2-O5 2.054(7), O1-C1 1.249(13), O2-C1 1.277(13), O3-C2 1.258(13), O4-C2 1.241(13), O5-C3 1.247(13), O6-C3 1.258(14); P1-Au1-Cl1 178.94(12), P2-Au2-Cl2 170.7(2), P3-Au3-Cl3 177.3(2), O3-Zn1-O1 93.1(3), O3-Zn1-O2 97.4(3), O3-Zn1-O6 95.9(3), O1-C1-O2 120.6(10), O4-C2-O3 127.2(10), O5-C3-O6 127.2(10).

3.1.4 Synthesis of a flexible zinc-metalloligand bearing terminal phosphines

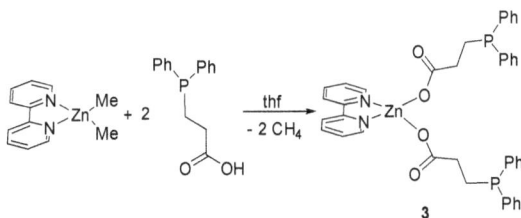

Scheme 3.1.3: Synthesis of complex **3**.

In order to make high nuclearity complexes of gold, aurophilic interactions can play a major role,[124,132] and flexible ligands can promote such metal-metal interactions.[133] We anticipated that, by using the flexible 3-(diphenylphosphino)propionic acid (H-L^{Et}) ligand, Au-Au interactions may be possible and result in the formation of high nuclearity Zn-Au complexes similar to **2**. The zinc-based metalloligand **3** with a flexible linker between the carboxylate and phosphine donor sites was synthesized by reaction of [(Bipy)ZnMe$_2$] with H-L^{Et} in 1:2 molar ratio. As a result, complex **3**, {(Bipy)Zn(O$_2$C(C$_2$H$_4$)PPh$_2$)$_2$}, was isolated in 62% yield (Scheme 3.1.3). The ^{1}H NMR spectrum shows multiplets from δ 2.32 to 2.43 ppm corresponding to the ethylene unit of the L^{Et} ligand. In the aromatic region, signals are observed from δ 7.25 to 8.98 ppm which correspond to the aryl groups of the L^{Et} and bipyridine ligands. In the ^{31}P{^{1}H} NMR spectrum, a single resonance is detected at δ -14.7 ppm. Single crystals of complex **3** were obtained by slow vapour diffusion of pentane into a thf solution of the complex. The compound crystallizes in the orthorhombic space group *Aea2* with half a molecule in the asymmetric unit cell. The Zn atom is in a distorted tetrahedral environment, coordinated by the two N donors of the chelating bipyridine and two carboxylate O atoms of two L^{Et} ligands (Figure 3.1.3). The Zn-N bond length (2.058(3) Å) in complex **3** is slightly shorter than the corresponding distances in complex **1** (2.0658(14) Å and 2.0872(12) Å). The Zn-O bond length (1.937(3) Å) is similar to that corresponding to the κ^1-coordinated carboxylate group in complex **1** (Zn1-O3: 1.921(2) Å). The

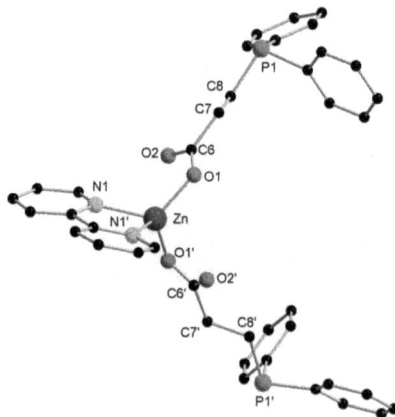

Figure 3.1.3: Molecular structure of complex **3**. Hydrogen atoms are omitted for clarity. Selected bond lengths (Å) and angles [°]: Zn-O1 1.937(3), Zn-N1 2.058(3), Zn-N1' 2.058(3), O1-C6 1.286(4), O2-C6 1.235(4); O1-Zn-O1' 118.6(2), O1-Zn-N1 106.26(10), N1-Zn-N1' 79.6(2), O2-C6-O1 122.6(3).

κ^1 coordinated carboxylate group in complex **3** features a O2-C6-O1 (122.6(3)°) angle intermediate between the O-C-O angle in complex **1** (κ^1 = 123.70(13)°, κ^1 = 121.66(13)°).

3.1.5 Synthesis of nonanuclear Zn-Au heterometallic complex featuring aurophilic interactions

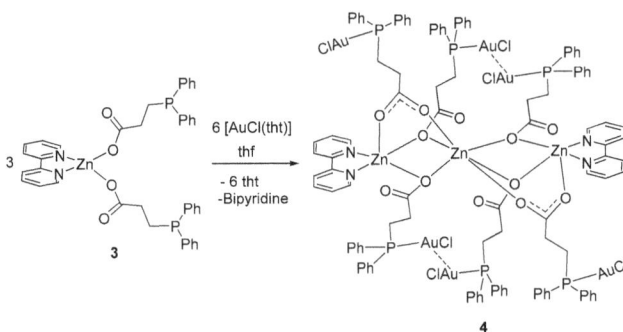

Scheme 3.1.4: Synthesis of complex **4**.

With the zinc-based metalloligand **3** in hand, its coordination chemistry towards gold(I) centres was studied. The reaction of **3** with [AuCl(tht)] furnished complex **4**, [(Bipy)$_2$Zn$_3$\{O$_2$C(C$_2$H$_4$)PPh-$_2$(AuCl)\}$_6$], isolated in 68% yield after a short workup (Scheme 3.1.4). In contrast to signals observed in complex **3**, the ^{1}H NMR spectrum of **4** shows two separate sets of multiplets at δ 2.54-2.60 and 2.73-2.79 ppm corresponding to the protons of the ethyl group of the L^{Et} ligand. As observed for complex **2a**, coordination of the phosphine to the AuCl moiety results in downfield shifts of the aromatic resonances in the ^{1}H NMR spectrum and of the phosphorous resonance in the ^{31}P\{^{1}H\} NMR spectrum (from δ -14.7 in **3** to δ 29.5 ppm in **4**). The ^{1}H NMR spectrum of complex **4** is apparent for the proposed structure (Scheme 3.1.4). The solid-state structure of **4** was unambiguously established by single-crystal X-ray analysis (Figure 3.1.4). Complex **4** crystallizes in the triclinic space group $P\bar{1}$ with half a molecule in the asymmetric unit cell. Similarly to **2**, a nonanuclear Zn$_3$Au$_6$ complex is formed *via* the loss of one bipyridine molecule. The three Zn atoms in complex **4** are arranged in a similar manner as in complex **2**. The terminal Zn1 and Zn1' are in a square pyramidal geometry, each coordinated by one chelating bipyridine ligand and three L^{Et} carboxylate moieties. The central Zn2 is in an octahedral coordination environment, only surrounded by carboxylate ligands. In complex **4**, the Zn1-N1 (2.139(6) Å) and Zn1-N2 (2.103(7) Å) bond distances are elongated as compared to the Zn-N bond length in complex **3** (2.058(3) Å). The oxygen atoms of the carboxylate group are in two different coordination modes, κ^1 and κ^2. The Zn-O bond lengths (from 1.989(5) to 2.174(5) Å) and the Au-P bond distances (2.227(2)-2.240(2) Å) are in the expected range. Interestingly, two intramolecular Au-Au interactions are observed with Au1-Au2 separation (3.3478(5) Å) in the typical range for aurophilic interactions (2.70-3.50 Å).[130a,131] The Au-Au interactions in complex **4** can be classified as semi-supported aurophilic interactions.[131b,131e] As observed in complex **2**, the N atoms of the bipyridine ligand and three Zn atoms in **4** are coplanar.

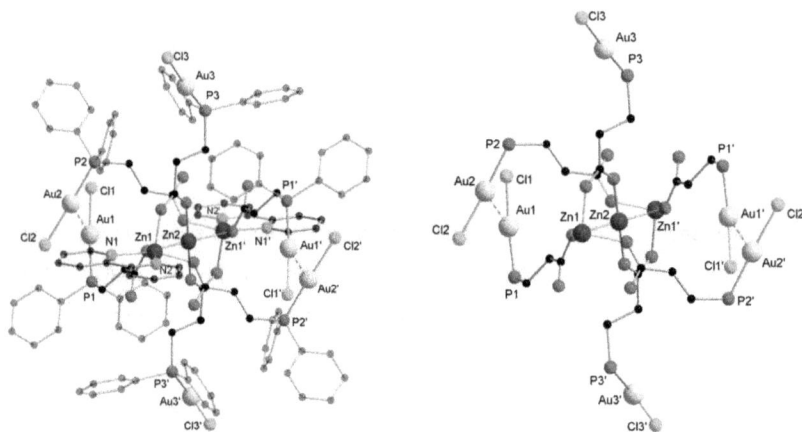

Figure 3.1.4: Molecular structure of complex **4** (left) and simplified view where the phenyl groups on the phosphine have been omitted (right). Hydrogen atoms are omitted for clarity. Selected bond lengths (Å) and angles [°] : Au1-Au2 3.3478(5), Au1-Cl1 2.285(2), Au2-Cl2 2.292(2), Au3-Cl3 2.279(2), Au1-P1 2.236(2), Au2-P2 2.240(2), Au3-P3 2.227(2), Zn1-N1 2.103(7), Zn1-N2 2.139(6), Zn1-O1 1.989(5), Zn1-O3 2.050(5), Zn1-O5 2.058(5), Zn2-O2 2.002(5), Zn2-O3 2.174(5), Zn2-O5 2.145(5), O1-C29 1.253(8), O2-C29 1.261(8), O3-C13 1.298(8), O4-C13 1.225(9), O5-C14 1.311(8), O6-C14 1.209(9); P1-Au1-Cl1 172.81(9), P2-Au2-Cl2 174.95(8), P3-Au3-Cl3 175.52(8), O6-C14-O5 124.7(6), O4-C13-O3 122.5(6), O1-C29-O2 125.7(6).

3.2 Lanthanide-TM carbonyl complexes

Complexes **8** to **12** in the following section has already been published:

R. Yadav, T. Simler, M. T. Gamer, R. Köppe, and P. W. Roesky, *Chem. Commun.* **2019**, *55*, 5765-5768. Schemes 3.2.4-3.2.7 and figures 3.2.7-3.2.16 are adopted from Ref. [149] with permission from the royal society of chemistry.

3.2.1 Introduction

In recent years, Ln-TM heterometallic complexes have attracted interest due to their photophysical properties,[134] magnetic properties,[135] as well as their catalytic activities.[136] Among these complexes, Ln-TM carbonyl complexes are major representatives due to an intrinsic interest in their bonding properties as well as their application in material science.[52] For example, Dy-TM carbonyl complexes have demonstrated single-molecule magnetic behaviour, which significantly depends on the nature of surrounding ligands and type of Dy-TM interaction *i.e.* presence or absence of a direct bond between Dy and TM.[55e,137] Different approaches to make Ln-TM carbonyl complexes have been discussed in Section 1.3.2.

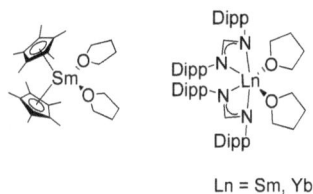

Ln = Sm, Yb

Chart 3.2.1: Divalent lanthanide complexes used in the present work for the reduction of TM carbonyl complexes.

Although the chemistry of Ln^{III}-TM carbonyl complexes has been well explored, no comparative study of the effect of the ligand environment around Ln centre has been done so far. Recently, the nature of the ligands was found to have significant effect on the outcome of redox reactions based on divalent lanthanides.[46a,50,84] Depending on the lanthanide metal (Sm or Yb), the surrounding ligands, and the choice of TM carbonyl, different types of Ln^{III} TM carbonyl complexes can be expected. Our plan is to examine the reactivity of mid-row TM carbonyls, covering examples from the 3d and 5d series, namely $[Fe_2(CO)_9]$, $[Fe_3(CO)_{12}]$, $[Co_2(CO)_8]$, $[Mn_2(CO)_{10}]$, and $[Re_2(CO)_{10}]$, with Ln^{II} (Ln = Yb and Sm) complexes. Two different types of ligands featuring distinct steric and electronic properties will be studied as supporting ligands for the Ln^{II} complexes. For that purpose, the sterically bulky N-donor based DippForm ligand and the less sterically demanding C-donor based Cp* ligand have been used in the coordination sphere of the lanthanide cation. The reactivity of $[Cp*_2Yb^{II}(OEt_2)_2]$ towards TM carbonyls has already been reported (Section 1.3.1). However, the reactivity of its samarium analogue, $[Cp*_2Sm^{II}(thf)_2]$, has not been examined for similar reactions. Therefore, in the present work, when using Cp* as a ligand, only the reactivity of $[Cp*_2Sm^{II}(thf)_2]$ will be investigated towards TM carbonyls, and the obtained results will be compared with the previous reports on $[Cp*_2Yb^{II}(OEt_2)_2]$.

3.2.2 Synthesis of Ln-TM carbonyl complexes by using bulky amidinates as ligands on Ln^{II}

The synthesis of $[(DippForm)_2Ln^{II}(thf)_2]$ (Ln = Sm,[138] Yb[49]) has been already reported by Junk and Deacon. The reaction between $[(DippForm)_2Sm^{II}(thf)_2]$ and one equivalent of $[Fe_2(CO)_9]$ or half an equivalent of $[Fe_3(CO)_{12}]$ in thf at 60 °C resulted in the formation of $[\{(DippForm)_2Sm^{III}\}_2\{(\mu_3\text{-}CO)_2(CO)_9Fe_3\}]$ (5) (Scheme 3.2.1). After a short work-up, red coloured crystals were isolated in 33% yield. The solid-state IR spectrum showed characteristic $\tilde{v}_{CO}$ resonances at 2011 (vs), 1979 (vs), 1967 (s), 1878 (m), 1830 (w), and 1696 (w) cm^{-1} (Figure 3.2.1). In the previously reported related complex $[\{(NEt_4)_2\}\{Fe_3(CO)_{11}\}]$,[139] similar $\tilde{v}_{CO}$ at 1938, 1910, 1890, and 1670 cm^{-1} were observed. The low-frequency stretch in the carbonyl region at 1696 cm^{-1} suggests the presence of bridging isocarbonyls between the Sm and Fe atoms.

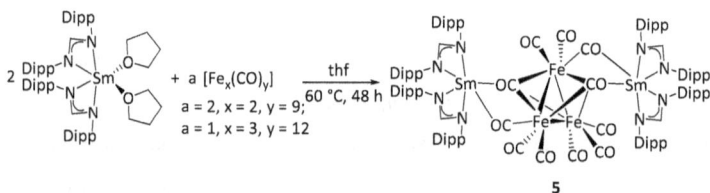

Scheme 3.2.1: Synthesis of complex 5.

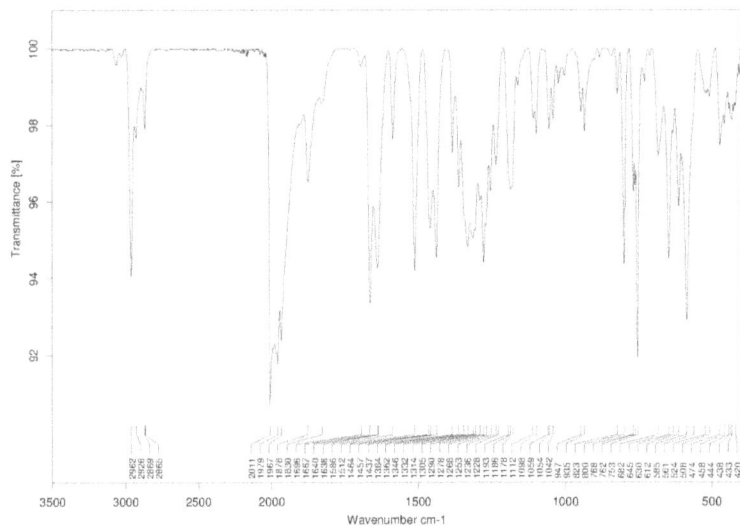

Figure 3.2.1: The solid-state IR spectrum of complex **5**.

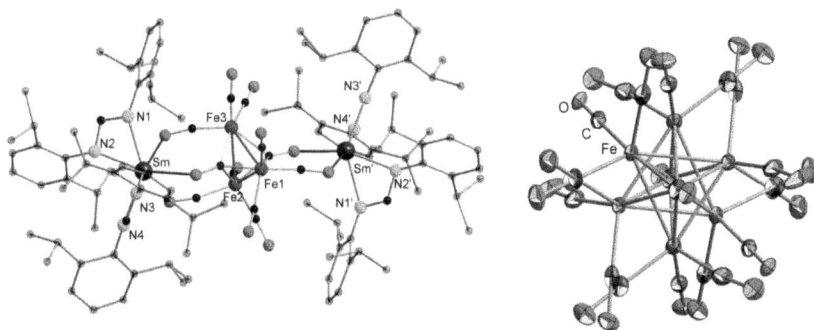

Figure 3.2.2: Molecular structure of **5** in the solid state (left) and a view of complete disorder of the central $[(\mu_3\text{-}CO)_2(CO)_9Fe_3]^{2-}$ moiety of complex **5** with thermal ellipsoids at the 40% probability with (right). H atoms are omitted for clarity. Selected bond distances (Å) and angles [°]: N1-Sm 2.409(2), N2-Sm 2.426(2), N3-Sm 2.381(2), N4-Sm 2.418(2), N1-C11 1.329(3), N2-C11 1.329(3), N4-C36 1.323(3), N3-C36 1.336(3); N1-C11-N2 117.87(2), N4-C36-N3 117.9(2), N1-Sm-N2 56.20(6), N1-Sm-N4 142.11(6), N3-Sm-N1 108.16(6), N3-Sm-N2 109.53(6), N3-Sm-N4 56.65(6). The bond lengths and angles for the $[(\mu_3\text{-}CO)_2(CO)_9Fe_3]^{2-}$ moiety cannot be described precisely due to the disorder.

The molecular structure of complex **5** was determined by single crystal X-ray diffraction studies. Complex **5** crystallizes in the monoclinic space group $P2_1/n$ with half a molecule in the asymmetric unit cell. The solid-state structure showed that the two [(DippForm)$_2$Sm^{III}]$^+$ moieties are connected to a [(μ_3-CO)$_2$(CO)$_9$Fe$_3$]$^{2-}$ anion through bridging isocarbonyls (Figure 3.2.2). In complex **5**, each Sm atom lies in a hexacoordinated environment, surrounded by four N atoms of bidentate amidinate ligands and two O atoms of isocarbonyl bridges. Interestingly, no coordinated thf molecule can be seen on the Sm atoms in complex **5**, possibly because of the recrystallization of the complex from hot toluene and also the steric crowding around the Sm atoms. The average Sm-N bond length (2.408(2) Å) in complex **5** is significantly shorter than in [(DippForm)$_2$Sm^{II}(thf)$_2$] (Sm-N(average) 2.573(3) Å).[138] This shortening can be explained by the decrease in the ionic radius of Sm when oxidized from the +2 to the +3 oxidation state.[140] The other bond lengths in complex **5** could not be discussed in detail as the central [(μ_3-CO)$_2$(CO)$_9$Fe$_3$]$^{2-}$ moiety is disordered over two positions. The formation of the [(μ_3-CO)$_2$(CO)$_9$Fe$_3$]$^{2-}$ moiety can be explained by two single-electron transfer steps from two molecules of [(DippForm)$_2$Sm^{II}(thf)$_2$] to one [Fe$_3$(CO)$_{12}$], accompanied by the loss of one CO ligand. In complex **5**, a triangular Fe core is sandwiched between two [(DippForm)$_2$Sm^{III}]$^+$ moieties by means of bridging isocarbonyls, which is in contrast to the product obtained in the previously reported reaction of [Fe$_3$(CO)$_{12}$] or [Fe$_2$(CO)$_9$] with [Cp*$_2$Yb^{II}(OEt$_2$)$_2$]. In the case of [Cp*$_2$Yb^{II}(OEt$_2$)$_2$], an isoelectronic [(μ_2-CO)$_4$(CO)$_7$Fe$_3$)]$^{2-}$ anion was obtained but the arrangement of the three Fe atoms was linear instead of triangular.[141] The possible reason for such a difference may be the sterically demanding nature of DippForm ligands that enforces the triangular arrangement of the Fe atom in the central core. Owing to the formation of

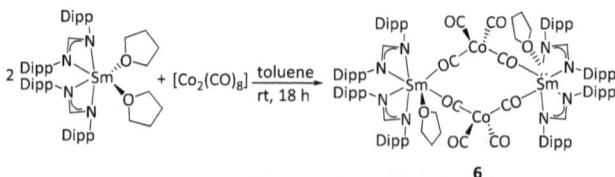

Scheme 3.2.2: Synthesis of complex 6.

unexpected products in the reaction with iron carbonyls, the reactivity of [(DippForm)$_2$Sm^{II}(thf)$_2$] was examined towards the neighbouring metal carbonyls. The reaction of [(DippForm)$_2$Sm^{II}(thf)$_2$] with half an equivalent of [Co$_2$(CO)$_8$] in toluene at room temperature resulted in the formation of [{(DippForm)$_2$Sm^{III}(thf)}$_2${(μ-CO)$_2$(CO)$_2$Co}$_2$] (6) (Scheme 3.2.2). Analytically pure yellow-coloured crystals of complex 6 were grown from slow evaporation of toluene in 56% yield. The solid-state IR spectrum of complex 6 showed characteristic $\tilde{\nu}_{CO}$ resonances at 2020 (m), 1951 (br), 1935 (br), 1922 (br), 1904 (br), 1842 (s), 1819 (s), and 1782 (s) cm^{-1} (Figure 3.2.3). The stretching frequency in the lower $\tilde{\nu}_{CO}$ region at 1782 (s) cm^{-1} can be assigned to bridging isocarbonyls between the Sm and Co atoms. The solid-state IR spectrum of complex 6 is comparable to that of a previously reported Yb-Co complex, [{(Cp*)$_2$Yb^{III}(thf)}{(μ-CO)(CO)$_3$Co}], which also showed similar low-frequency stretches at 1798 (m) and 1761 (s) cm^{-1} due to the presence of an isocarbonyl bridge.[142] The four terminal $\tilde{\nu}_{CO}$ from 2019 to 1819 cm^{-1} and the $\tilde{\nu}_{CO}$ in the lower region at 1782 cm^{-1} are in agreement with the proposed structure of complex 6 (Scheme 3.2.2). The molecular structure of complex 6 was determined by single crystal X-ray diffraction studies (Figure 3.2.4). Complex 6 crystallizes in the monoclinic space group $P2_1/n$ with half a molecule in the asymmetric unit cell. As for the formation of 5, the formation of complex 6 can be rationalised by single electron transfer steps from two [(DippForm)$_2$Sm^{II}(thf)$_2$] molecules to one molecule of [Co$_2$(CO)$_8$], resulting in the homolytic cleavage of the Co-Co bond and the formation of two [Co(CO)$_4$]$^-$ anions along with two [(DippForm)$_2$Sm^{III}(thf)]$^+$ cations. The two [(DippForm)$_2$Sm^{III}(thf)]$^+$ moieties are linked to the two [Co(CO)$_4$]$^-$ anions through isocarbonyl bridges. In complex 6, each Sm atom is heptacoordinated, surrounded by four N atoms of two bidentate amidinate ligands, two O atoms of two bridging isocarbonyls, and one O atom of a coordinated thf. The average Sm-N bond length (2.434(3) Å) in complex 6 is shorter than that in [(DippForm)$_2$Sm^{II}(thf)$_2$] (2.573(3) Å),[48] which can be attributed to the decrease in the ionic radius of Sm upon going in the +3 oxidation state.[140] A further evidence of the +3 oxidation state of Sm in complex 6 lies in the shortening of the Sm-O1(thf) bond length from 2.560(3) Å in [(DippForm)$_2$Sm^{II}(thf)$_2$] to 2.422(2) Å in complex 6. The tetracarbonylcobaltate anion has a distorted tetrahedral geometry with C-Co-C angles ranging from 105.8(2)° to 116.1(2)°.

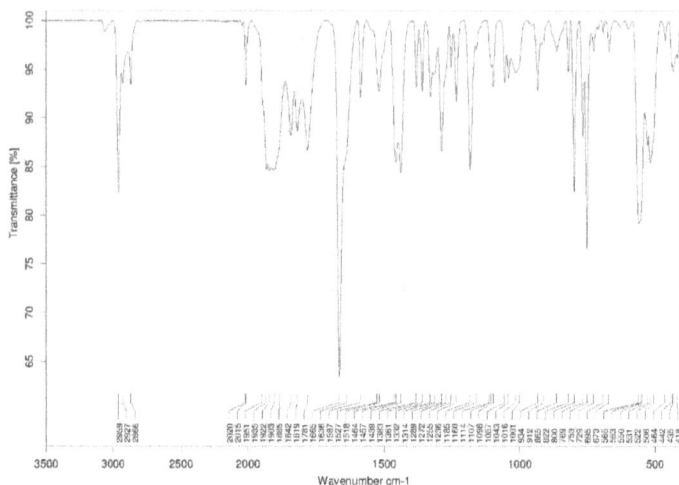

Figure 3.2.3: The solid-state IR spectrum of complex **6**.

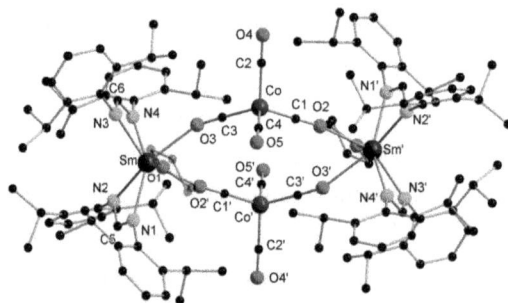

Figure 3.2.4: Molecular structure of **6**. Hydrogen atoms are omitted for clarity. Selected bond distances (Å) and angles [°]: Sm-O1 2.422(2), Sm-O2′ 2.488(2), Sm-O3 2.552(3), Sm-N1 2.431(3), Sm-N2 2.449(3), Sm-N3 2.409(3), Sm-N4 2.449(3), Co-C1 1.733(4), Co-C2 1.787(4), Co-C3 1.747(4), Co-C4 1.782(4), O2-C1 1.175(4), O3-C3 1.170(4), O4-C2 1.138(5), O5-C4 1.145(5), N1-C5 1.321(4), N2-C5 1.318(4), N3-C6 1.330(5), N4-C6 1.326(5); O2′-Sm-O3 67.28(8), N1-Sm-N2 55.35(9), N3-Sm-N4 56.61(9), N2-C5-N1 118.4(3), N4-C6-N3 120.3(3), C1-Co-C2 108.8(2), C1-Co-C3 116.1(2), C1-Co-C4 110.1(2), C3-Co-C2 109.7(2), C3-Co-C4 105.8(2), C4-Co-C2 105.8(2).

34

The O2-C1 (1.175(4) Å) and O3-C3 (1.170(4) Å) bond distances involving the bridging isocarbonyls are longer than the O4-C2 (1.138(5) Å) and O5-C4 (1.145(5) Å) bond lengths due to the coordination of O2 and O3 to the [(DippForm)$_2$Sm^{III}(thf)]$^+$ moiety. A further effect of the formation of bridging isocarbonyls can be seen by analysis of the Co-C bond lengths: the Co-C(bridging) bonds are shortened and strengthened as compared to the Co-C(terminal) bonds (Co-C1 (1.733(4) Å) and Co-C3 (1.747(4) Å) vs Co-C4 (1.782(4) Å) and Co-C2 (1.787(4) Å), respectively). To the best of our knowledge, no Sm^{III}-Co carbonyl complex featuring an isocarbonyl bridging between Sm and Co has been reported to date.[143] Evans et al. has reported the reduction of [Co$_2$(CO)$_8$] by [Cp*$_2$Sm^{II}(thf)$_2$] but without any solid-state structure. However, in the reaction between [Sm^{II}I$_2$(thf)$_2$] and [Co$_2$(CO)$_8$] in thf, the solvent separated ionic pair, [Sm^{III}I$_2$(thf)$_4$][Co(CO)$_4$], was obtained and structurally characterized.[144]

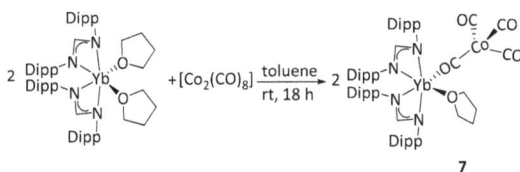

Scheme 3.2.3: Synthesis of complex **7**.

The reaction of half an equivalent of [Co$_2$(CO)$_8$] with [(DippForm)$_2$Yb^{II}(thf)$_2$] in toluene at room temperature resulted in the formation of complex [{(DippForm)$_2$Yb^{III}(thf)}{(μ-CO)(CO)$_3$Co}] (**7**) (Scheme 3.2.3). After a short work-up, orange-coloured crystals of complex **7** were isolated in 63% yield. The solid-state IR spectrum of complex **6** shows characteristic $\tilde{v}_{CO}$ stretches from 2031 to 1748 cm^{-1} (Figure 3.2.5). These stretches are similar to those in [{(Cp*)$_2$Yb^{III}(thf)}{(μ-CO)(CO)$_3$Co}] (2023 to 1761 cm^{-1}).[55a] The low-frequency band at 1748 cm^{-1} in the $\tilde{v}_{CO}$ stretching region in complex **7** indicates the presence of bridging isocarbonyls. The latter can further be seen in the solid-state structure of complex **7**. Complex **7** crystallizes in the triclinic space group $P\bar{1}$ with one molecule in the asymmetric unit cell. The molecular structure showed that complex **7** crystallizes in a monomeric form in contrast to the dimeric arrangement of complex **6** (Figure 3.2.6). The possible reason for this difference could be a combined effect of the smaller ionic

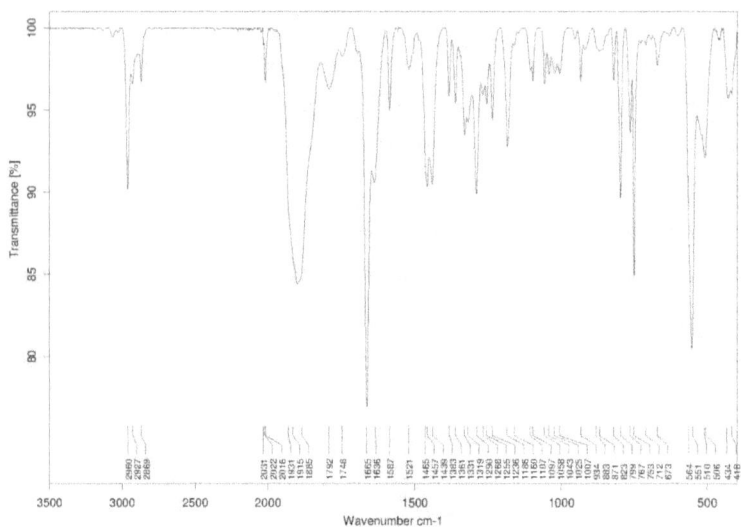

Figure 3.2.5: The solid-state IR spectrum of complex **7**.

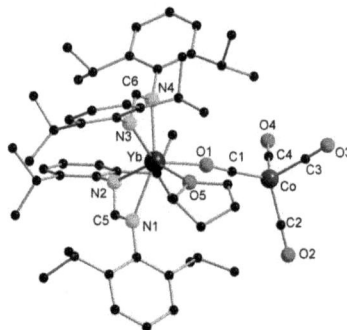

Figure 3.2.6: The molecular structure of **7** in solid-state. Hydrogen atoms are omitted for clarity. Selected bond distances (Å) and angles [°]: Yb-O1 2.248(2), Yb-O5 2.304(2), Yb-N1 2.337(2), Yb-N2 2.307(2), Yb-N3 2.292(2), Yb-N4 2.324(2), Co-C3 1.705(3), Co-C4 1.787(4), Co-C5 1.764(3), Co-C6 1.783(5), O1-C3 1.199(3), O2-C4 1.153(5), O3-C5 1.139(4), O4-C6 1.144(5), N1-C1 1.314(3), N2-C1 1.316(3), N3-C2 1.319(3), N4-C2 1.314(3); O1-Yb-O5 81.70(8), N2-Yb-N1 58.23(7), N3-Yb-N4 58.68(8), N1-C1-N2 118.4(2), N4-C2-N3 118.4(2), C3-Co-C4 115.8(2), C3-Co-C5 109.6(2), C3-Co-C6 111.8(2), C5-Co-C4 106.8(2), C5-Co-C6 104.6(2), C6-Co-C4 107.5(2).

radius of Yb^{3+} as compared to Sm^{3+} and the sterically demanding nature of the DippForm ligands. The Yb centre is in a distorted octahedral environment, coordinated by four N atoms of two bidentate amidinate ligands, one O atom of an isocarbonyl group, and one O atom of a coordinated thf molecule. In complex 7, the Yb-N bond distances involving the amidinate ligands (2.325(2) Å, average) and the Yb-O5 bond length (2.304(2) Å) are shortened in comparison to [(DippForm)$_2$Yb^{II}(thf)$_2$][49] (Yb-N (2.447 Å) and Yb-O (2.461 Å), average), consistent with the oxidation of Yb from the +2 to the +3 oxidation state.[140] Similarly to 6, the tetracarbonylcobaltate moiety in 7 has a distorted tetrahedral coordination environment with C-Co-C angles varying from 104.6(2)° to 115.8(2)°. The Co-C3 (1.705(3) Å) bond length in 7 is almost the same as the Co-C(bridged) (1.699(3) Å) bond distance in [{Cp*$_2$Yb^{III}}(thf)(μ-CO)((CO)$_3$Co]. The average Co-C(terminal) bond length in 7 and [{Cp*$_2$Yb^{III}}(thf)(μ-CO)(CO)$_3$Co] are also similar (1.77 Å (average) in both cases).[55a]

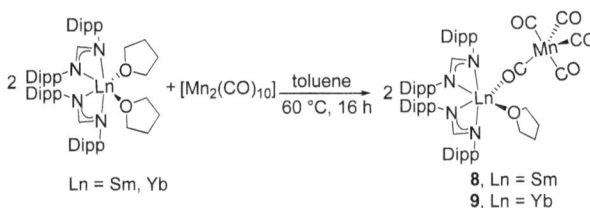

Scheme 3.2.4: Synthesis of complex 8 and 9.

The reactivity of [(DippForm)$_2$Ln^{II}(thf)$_2$] (Ln = Sm and Yb) was then examined towards [Mn$_2$(CO)$_{10}$]. The reaction of [(DippForm)$_2$Ln^{II}(thf)$_2$] with half an equivalent of [Mn$_2$(CO)$_{10}$] in toluene at 60 °C resulted in the formation [{(DippForm)$_2$Ln^{III}(thf)}{(μ-CO)(CO)$_4$Mn}] (Ln = Sm (8) and Yb (9)) (Scheme 3.2.4). The presence of low-frequency stretches at 1734 cm^{-1} (vw) for 8 and 1701 (w) cm^{-1} for 9 are indicating bridging isocarbonyls (Figure 3.2.7 and 3.2.8, respectively).[55c,62] The molecular structures of complexes 8 and 9 were determined by X-ray crystallography (Figure 3.2.9). The solid-state structures of 8 and 9 show that both compounds crystallize in the triclinic space group $P\bar{1}$, with isostructural arrangements. Therefore, only the structure of complex 8 is discussed here. Selected bond lengths for both 8 and 9 are given in the caption of figure 3.2.9.

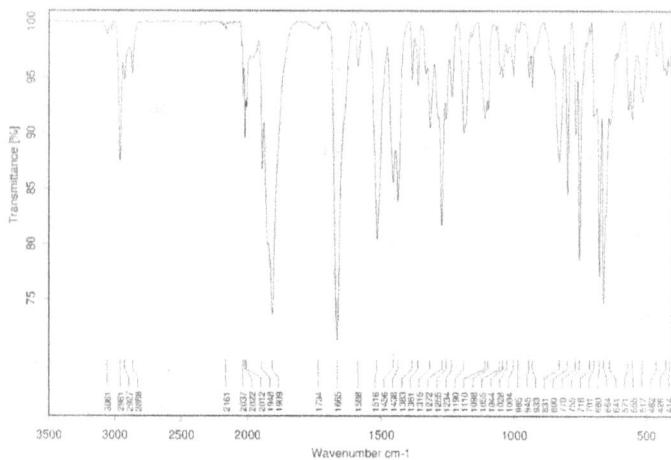

Figure 3.2.7: The solid-state IR spectrum of complex **8**.

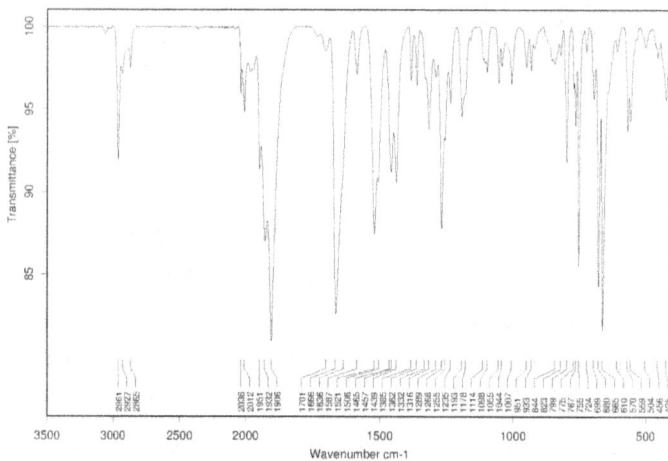

Figure 3.2.8: The solid-state IR spectrum of complex **9**.

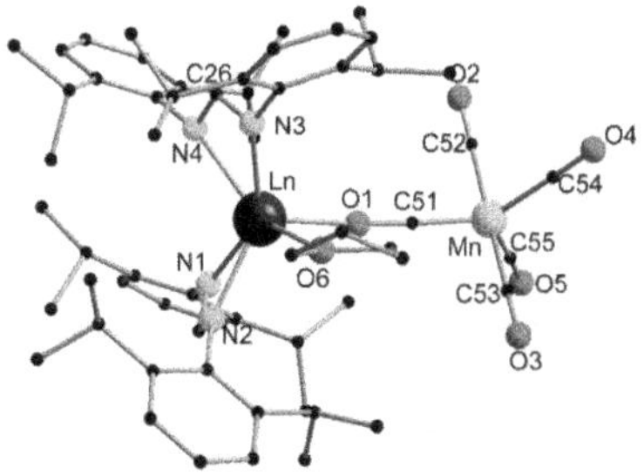

Figure 3.2.9: Molecular structures of **8** (Ln = Sm) and **9** (Ln = Yb) in the solid state. H atoms and non-coordinating solvent molecules are omitted for clarity. Selected bond distances (Å) and angles [°]: For **8**: Sm-O1 2.325(2), Sm-O6 2.435(2), Sm-N1 2.474(2), Sm-N2 2.344(2), Sm-N3 2.418(2), Sm-N4 2.399(2), Mn-C51 1.739(3), Mn-C52 1.823(4), Mn-C53 1.823(4), Mn-C54 1.826(3), Mn-C55 1.820(3), O1-C51 1.205(3), O2-C52 1.148(4), O3-C53 1.150(4), O4-C54 1.150(4), O5-C55 1.152(3); N2-Sm-N1 56.06(6), N4-Sm-N3 55.78(6), N1-C1-N2 117.6(2), C51-Mn-C52 87.72(13), C51-Mn-C53 91.30(14), C51-Mn-C54 128.16(13), C51-Mn-C55 117.13(13), C52-Mn-C54 87.5(2), C53-Mn-C52 174.95(14), C53-Mn-C54 89.2(2), C55-Mn-C52 93.11(14), C55-Mn-C53 91.7(2), C55-Mn-C54 114.67(13). For **9**: Yb-O1 2.192(2), Yb-O6 2.308(2), Yb-N1 2.280(2), Yb-N2 2.342(2), Yb-N3 2.300(2), Yb-N4 2.348(2), Mn-C51 1.725(3), Mn-C52 1.829(3), Mn-C53 1.835(3), Mn-C54 1.824(3), Mn-C55 1.831(3), O1-C51 1.211(3), O2-C52 1.148(3), O3-C53 1.138(3), O4-C54 1.149(4), O5-C55 1.147(3); N1-Yb-N2 58.85(7), N3-Yb-N4 57.85(7), C1-N1-C2 117.2(2), C51-Mn-C52 89.71(12), C51-Mn-C53 91.58(12), C51-Mn-C54 118.61(13), C51-Mn-C55 120.04(13), C52-Mn-C53 178.53(13), C52-Mn-C55 90.63(12), C54-Mn-C52 90.03(13), C54-Mn-C53 89.96(13), C54-Mn-C55 121.34(13), C55-Mn-C53 88.12(12).

In the solid-state structure of **8** (Figure 3.2.9), each Sm atom is in a distorted octahedral environment, coordinated by two amidinate ligands, one thf ligand and the oxygen atom of the bridging isocarbonyl. As expected, the average Sm-N bond distance (2.409 Å) agrees with the +3 oxidation state of the Sm atom. The [Mn(CO)$_5$]$^-$ fragment is in a trigonal bipyramidal geometry. Due to coordination to the [(DippForm)$_2$Sm^{III}(thf)]$^+$ moiety, the C51-O1 bond is weakened with a bond distance of 1.205(3) Å, as compared to the C-O (terminal) bond (1.150 Å (average)). Accordingly, the Mn-C51 bond gets strengthened by this interaction, with a Mn-C51 bond distance of 1.739(3) Å, shorter than the Mn-C(terminal) analogues (1.823 Å (average)). Formation of **8** and **9** occurred through cleavage of the Mn-Mn bond *via* a single electron transfer step. The [(DippForm)$_2$Sm^{III}(thf)]$^+$ and [Mn(CO)$_5$]$^-$ fragments that are generated further assemble through the formation of bridging isocarbonyls.

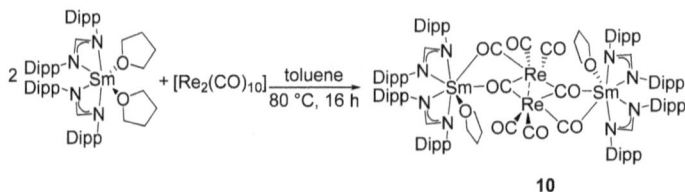

Scheme 3.2.5: Synthesis of complex **10** (simplified view).

The reactivity of 5d-TM carbonyls towards divalent lanthanides is far less studied than that of the 3d analogues.[145] After screening the reactivity of divalent lanthanides with a series of 3d TM carbonyl complexes (*vide supra*), we aimed to examine the behaviour of divalent lanthanides with a 5d-TM carbonyl complex, $[Re_2(CO)_{10}]$. The reaction of $[(DippForm)_2Sm^{II}(thf)_2]$ with half an equivalent of $[Re_2(CO)_{10}]$ in toluene at 80 °C resulted in the formation of the unprecedented complex $[\{(DippForm)_2Sm^{III}(thf)\}_2\{(\mu\text{-}\eta^2\text{-}CO)_2(\mu\text{-}\eta^1\text{-}CO)_2(CO)_4Re_2\}]$ (**10**) (Scheme 3.2.5). Red coloured crystals of complex **10** were grown by slow evaporation of toluene in 62% yield. The solid-state IR spectrum of complex **10** showed characteristic $\tilde{v}_{CO}$ resonances at 2070 (w), 2012 (m), 1973 (s), 1903 (m), and 1804 (w) cm^{-1} (Figure 3.2.11). The low-frequency stretch at 1804 cm^{-1} is consistent with the presence of bridging isocarbonyls.[146] The molecular structure of complex **10** was determined by X-ray crystallography (Figure 3.2.10). Each samarium atom is heptacoordinated, surrounded by two bidentate amidinate ligands, two oxygen atom of two bridging isocarbonyls, and one coordinating thf. In the $[Re_2(CO)_8]^{2-}$ anion, each rhenium atom is coordinated by the carbon donor of two terminal and three bridging CO ligands. The Re-C51 (μ-η^2-CO) bonds (2.049(3) and 2.206(3) Å) are longer than other Re-C bonds due to η^2 type bridging with two Re atoms. The Re-Re' bond length is 2.689(3) Å, which is substantially shorter than the distance of the Re-Re single bond in $[Re_2(CO)_{10}]$ (3.041 (11) Å),[147] suggesting a double-bond character between the two Re atoms. Besides, the $[Re_2(CO)_8]^{2-}$ anion fits with a 36 electron count only after considering a Re=Re double bond. These two factors support the presence of a double bond between the two Re atoms in complex **10**. However, the theoretical calculations support only a single bond (*vide infra*).

Formation of **10** presumably occurs through the transfer of 2e⁻ from two molecules of [(DippForm)$_2$Sm^{II}(thf)$_2$] to one molecule of [Re$_2$(CO)$_{10}$] in two SET steps, accompanied by the loss of two CO per [Re$_2$(CO)$_{10}$]. In complex **10**, the [Re$_2$(CO)$_8$]$^{2-}$ anion is entrapped between two [(DippForm)$_2$Sm^{III}(thf)]$^+$ moieties. Interestingly, even after the 2e⁻ reduction of [Re$_2$(CO)$_{10}$], the bond between the two rhenium atoms is retained. In contrast, single electron reduction of [Re$_2$(CO)$_{10}$] *e.g.* reduction of [Re$_2$(CO)$_{10}$] with alkali metals leads to the formation of [MRe(CO)$_5$] (M = alkali metal).[148] To best of our knowledge, the [Re$_2$(CO)$_8$]$^{2-}$ anion has never been reported before.[149] Recently, its lighter analogue, [Mn$_2$(CO)$_8$]$^{2-}$, has been isolated by reduction of [Mn$_2$(CO)$_{10}$] with silylene but the authors failed to isolate the [Re$_2$(CO)$_8$]$^{2-}$ anion under similar conditions.[150] The formation of the [(μ-CO)$_4$(CO)$_4$Re$_2$]$^{2-}$ fragment in compound **10** can be attributed to the high reductive nature of divalent samarium complexes and the sterically demanding nature of the [(DippForm)$_2$Sm^{III}(thf)]$^+$ moieties. The reaction of [(DippForm)$_2$Yb^{II}(thf)$_2$] with [Re$_2$(CO)$_{10}$] was also carried out under similar conditions, however, only starting materials were isolated. The possible reason for this lack of reactivity could be the lower reductive ability of divalent Yb complexes as compared to divalent Sm complexes.

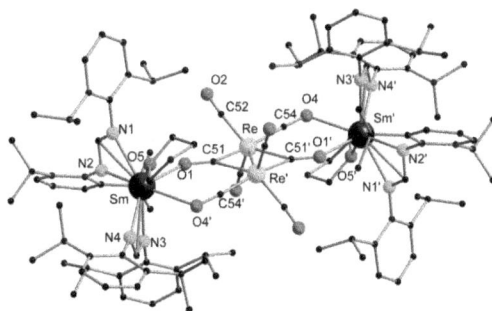

Figure 3.2.10: The molecular structure of **10** in the solid state. H atoms and non-coordinating solvent molecules are omitted for clarity. Selected bond distances (Å) and angles [°]: Sm-O1 2.365(2), Sm-O4' 2.628(2), Sm-O5 2.429(2), Sm-N1 2.495(3), Sm-N2 2.418(3), Sm-N3 2.454(3), Sm-N4 2.468(3), Re-Re' 2.689(3), Re'-C51 2.206(3), Re-C51 2.049(3), Re-C52 1.939(4), Re-C54 1.910(3), O1-C51 1.240(4), O2-C52 1.145(5), O4-C54 1.168(4); O1-Sm-O4' 67.85(7), N2-Sm-N1 55.48(9), N2-Sm-N3 90.76(9), N2-Sm-N4 114.70(9), N3-Sm-N4 55.67(8), N1-C1-N2 119.6(3), C54-Re-C52 90.4(2), Re-C51-Re' 78.30(11), C54-Re-C51 70.33(12).

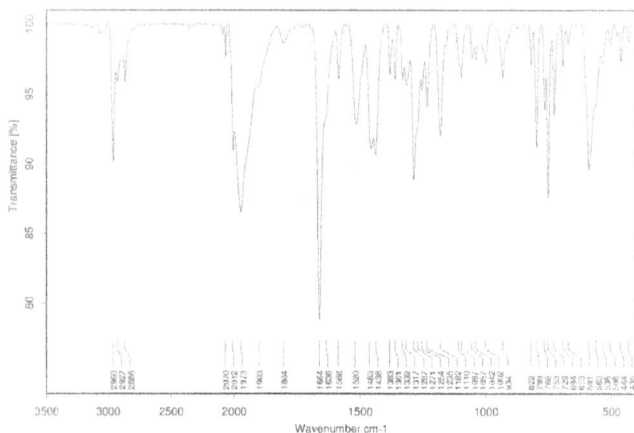

Figure 3.2.11: The solid-state IR spectrum of complex 10.

3.2.3 Synthesis of Ln-TM carbonyl complexes by using cyclopentadienyl as a ligand on Ln^{II}

Recently, Roesky *et al.* have shown that using ligands with different electronic properties and steric protection in the coordination sphere of divalent lanthanides, different activation of main-group elements and complexes can be achieved (Section 1.3.1).[46a,50,84,149] In order to understand if this observation is also valid for the reduction of transition metal carbonyls, the cyclopentadienyl based divalent samarium reagent, [(Cp*)$_2$Sm^{II}(thf)$_2$],[32] was employed.

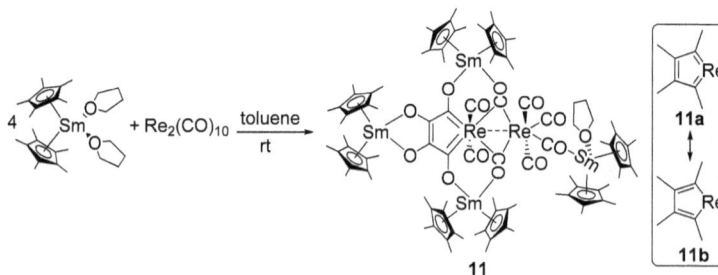

Scheme 3.2.6: Synthesis of complex 11 (simplified view). Possible isomers of 11 are double Fischer-carbene (11a) and metallacyclopentadiene (11b).

The reaction between [(Cp*)$_2$Sm^{II}(thf)$_2$] and [Re$_2$(CO)$_{10}$] in toluene at room temperature resulted in the formation of [{(Cp*)$_2$Sm^{III}}$_3${(μ-O$_4$C$_4$)(μ-η^2-CO)$_2$(μ-η^1-CO)(CO)$_5$Re$_2$}Sm^{III}(Cp*)$_2$(thf)] (11) as red-coloured single crystals isolated in 18% yield (Scheme 3.2.6). In the formation 11, Sm^{II} acts as a SET reagent leading to a central [(μ-O$_4$C$_4$)(μ-η^2-CO)$_2$(μ-η^1-CO)(CO)$_5$Re$_2$]$^{4-}$ core formed by a four-fold reduction process.[151] The solid-state IR spectrum of 11 is different from that of the starting material [Re$_2$(CO)$_{10}$] and shows characteristic $\tilde{v}_{CO}$ stretches at 2091 (vw), 2038 (w), 2010 (m), 1978 (vs), 1890 (m), 1858 (s), 1792 (s), 1777 (m), and 1733 (s) cm^{-1} (Figure 3.2.12). The low-frequency stretches at 1792 (s) and 1733 (s) cm^{-1} may be assigned to the bridging isocarbonyls.[152] The molecular structure of 11 was established by X-ray crystallographic studies. Complex 11 crystallizes in the monoclinic space group $P2_1/n$ with one molecule in the asymmetric unit cell. The solid-state structure of 11 revealed the formation of a hexametallic complex consisting of one [(μ-O$_4$C$_4$)(μ-η^2-CO)$_2$(μ-η^1-CO)(CO)$_5$Re$_2$]$^{4-}$, one [(Cp*)$_2$Sm^{III}(thf)]$^+$, and three [(Cp*)$_2$Sm^{III}]$^+$ moieties (Figure 3.2.13). The Sm1, Sm2, and Sm3 atoms are coordinated by two η^5-Cp* ligands and two oxygen atoms from bridging isocarbonyls or the newly formed μ-O$_4$C$_4$ entity. In contrast, Sm4 is coordinated by two η^5-Cp* ligands, one oxygen from a bridging isocarbonyl, and one oxygen from a coordinated thf. In accordance with the +3 oxidation state of Sm in 11, the average Sm-C(Cp*) bond length (2.704 Å) is significantly shorter than in [(Cp*)$_2$Sm^{II}(thf)$_2$] (2.860(3) Å),[32] due to the decreased ionic radius of Sm^{III} compared to Sm^{II}.[140] Both Re atoms are in an octahedral coordination environment and additionally bonded to each other. The Re-Re bond length in complex 11 (2.934(3) Å) is slightly shorter than the Re-Re single bond in [Re$_2$(CO)$_{10}$], (3.041(11) Å).[147] This difference suggests the presence of a single bond between two Re atoms in complex 11, however, theoretical calculations only suggest a weaker interaction (*vide infra*). Re1 and Re2 have different coordination environments. Re2 is coordinated to six CO ligands and Re1. On the other hand, Re1 is bonded to C1 and C4, two terminal CO, and two isocarbonyls (μ-η^2)-CO. The bond distances between Re1-C1 (2.188(6) Å) and Re1-C4 (2.164(7) Å) are in the typical range of conjugated rhenacyclic Fischer-type carbenes.[152] Two different possible forms can describe the bonding situation in the 5-membered rhenacycle: either a double Fischer-carbene (11a) or a metallacyclopentadiene (11b) (Scheme 3.2.6). The short C2-C3 bond distance in the central C$_4$O$_4$ fragment (1.406(9) Å (C2-C3) *vs.* 1.445(9) Å (C1-C2) and 1.460(9) Å (C3-C4)) suggests the higher weight of **11a**. Thus, compound **11** can be considered as a cyclic double Fischer-carbene type complex. The formation of the five membered Fischer-type rhenacycle in complex **11** occurred through an unprecedented tetramerization of CO ligands, by reductive C-C coupling. Interestingly, a total of 12 C-O units are present in the $[(\mu\text{-}O_4C_4)(\mu\text{-}\eta^2\text{-}CO)_2(\mu\text{-}\eta^1\text{-}CO)(CO)_5Re_2]^{4-}$ moiety, more than in the starting material $[Re_2(CO)_{10}]$. Since no external source of CO was present, the formation of **11** implies that more than one equivalent of $[Re_2(CO)_{10}]$ reacts with four equivalents of $[(Cp^*)_2Sm^{II}(thf)_2]$. During the course of the reaction, formation of a light green coloured precipitate was observed but unfortunately the latter could not be characterized. Owing to the non-stoichiometric nature of the reaction and the formation of unidentified side-products, the mechanism leading to the formation of complex **11** could not be established with certainty. To the best of our knowledge, CO tetramerization has never been observed with rhenium carbonyls. In addition, of all the Ln–TM carbonyl complexes reported earlier, none of them have shown reductive C-C coupling of CO ligands on TM carbonyls.[55a-c,56b,59,62] Although two reports describe a similar tetramerization of CO ligands, both were obtained by reactions between trimethylsilylhalide (halide = Br, or I)[153] or $[\{(Me_3Si)_2N\}BBr_2]^{[154]}$ and $[Na_2Fe(CO)_4]$. Furthermore, the resulting products did not feature Fischer-carbene type character but a metallcyclopentadiene type character (Scheme 3.2.6, **11b**). The reaction between trimethylsilylhalide (halide = Br, or I) or $[\{(Me_3Si)_2N\}BBr_2]$ and $[Na_2Fe(CO)_4]$ can be seen as a reaction of reduced TM carbonyl complex with Lewis acidic B or Si compounds. Inspired by this observation, the formation of complex **11** may involve the reduction of $[Re_2(CO)_{10}]$ by $[Cp^*_2Sm^{II}(thf)_2]$, generating reduced Re species which might perform CO tetramerization in the presence of the Lewis acidic $[Cp^*_2Sm^{III}]^+$ moieties.

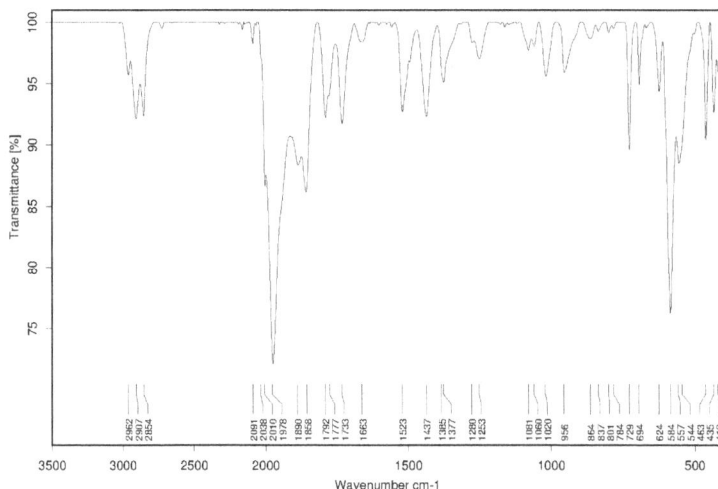

Figure 3.2.12: The solid-state IR spectrum of complex **11**.

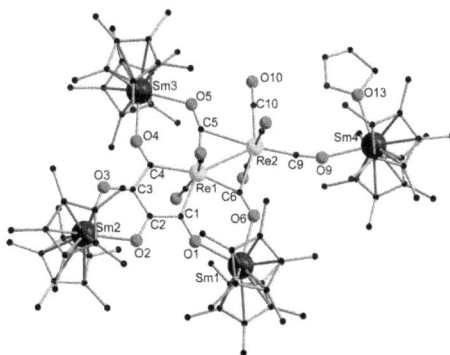

Figure 3.2.13: Molecular structure of **11** in the solid state. H atoms and non-coordinating solvent molecules are omitted for clarity. Selected bond distances (Å) and angles [°]: Sm1-O1 2.277(4), Sm1-O6 2.396(4), Sm2-O2 2.290(4), Sm2-O3 2.326(4), Sm3-O4 2.278(4), Sm3-O5 2.413(5), Sm4-O9 2.422(5), Sm4-O13 2.421(5), Re1-Re2 2.934(3), Re1-C1 2.188(6), Re1-C4 2.164(7), Re1-C6 2.052(6), Re1-C5 1.972(7), Re2-C5 2.547(6), Re2-C6 2.265(6), Re2-C9 1.878(7), O1-C1 1.257(7), O2-C2 1.314(7), O3-C3 1.294(8), O4-C4 1.261(8), O5-C5 1.192(8), O6-C6 1.209(8), O9-C9 1.179(8), C1-C2 1.445(9), C2-C3 1.406(9), C3-C4 1.460(9); C4-Re1-C1 75.0(2), Re1-C6-Re2 85.5(2), Re1-C5-Re2 79.9(2), C2-C1-Re1 116.1(4), C3-C2-C1 115.9(5), C2-C3-C4 114.8(6), C3-C4-Re1 117.2(4).

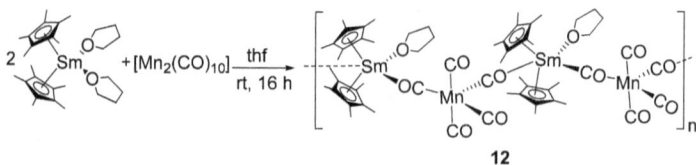

12

Scheme 3.2.7: Synthesis of a polymeric complex **12**.

Stimulated by the precedent results, we investigated whether a similar reactivity would also be observed between $[Mn_2(CO)_{10}]$ and $[Cp*_2Sm^{II}(thf)_2]$. The reaction of $[(Cp*)_2Sm^{II}(thf)_2]$ with half an equivalent of $[Mn_2(CO)_{10}]$ in toluene resulted in the formation of $[\{(Cp*)_2Sm^{III}(thf)\}\{(\mu\text{-}CO)_2(CO)_3Mn\}]_n$ (**12**) as red precipitate (Scheme 3.2.7). The solid-state IR spectrum of complex **12** shows characteristic $\tilde{\nu}_{CO}$ stretches at 2013 (m), 1968 (s), 1940 (s), 1875 (w), 1830 (s), 1770 (vs), 1744 (s) cm^{-1}. The low-frequency stretches at 1770 and 1744 cm^{-1} can be assigned to the bridging isocarbonyls (Figure 3.2.14),[55c,62] which was confirmed by analysis of the X-ray crystal structure of complex **12** (Figure 3.2.15). Red crystals of **12** suitable for X-ray diffraction analysis were grown by slow cooling of a hot thf solution of the complex. Complex **12** crystallizes in the triclinic space group $P\bar{1}$ as a polymeric complex. Each samarium atom is surrounded by two Cp* ligands, the oxygen atom of a thf ligand, and two oxygen atoms of bridging isocarbonyls. The average Sm-C (Cp*) bond distance (2.723 Å) is shorter than in $[(Cp*)_2Sm^{II}(thf)_2]$ (2.86(3) Å).[32] The $[Mn(CO)_5]^-$ anionic fragment has a trigonal bipyramidal geometry with two bridging CO and three terminal CO. The average bond lengths of Mn-C 1.834 Å (terminal) and Mn-C 1.774 Å (bridged) in complex **12** are similar as those in $[\{Cp*_2Yb^{III}\}\{(\mu\text{-}CO)_3\{Mn(CO)_2\}]_n$, Mn-C 1.820(5) Å (terminal, average) and Mn-C 1.791(13) Å (bridged, average).[155] Given the high reduction potential of $[(Cp*)_2Sm^{II}(thf)_2]$,[34] a Mn-Mn bond cleavage upon reaction of $[Mn_2(CO)_{10}]$ with $[(Cp*)_2Sm^{II}(thf)_2]$ is expected, resulting in the formation of a $[Mn(CO)_5]^-$ anion, which combined with the $[(Cp*)_2Sm^{III}(thf)]^+$ cation to generate a polymeric chain. Notably, polymerization only occurs for complex **12**, while, in case of complexes **8** and **9**, only monomeric complexes of Ln-Mn carbonyl were isolated. This reactivity difference could be rationalised in term of steric bulk around the lanthanide metal. The steric crowding exerted by the two bulky amidinate DippForm ligands in the case complexes **8** and **9** forces these complexes to stay in a monomeric form despite the hexacoordination of Sm (**8**) and Yb (**9**).

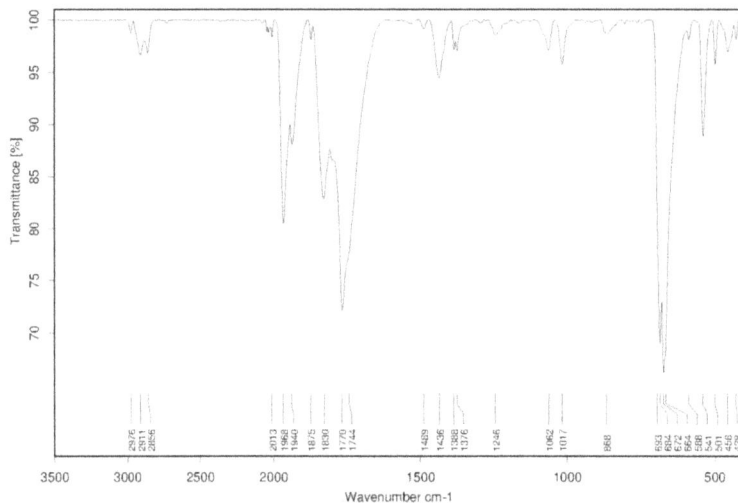

Figure 3.2.14: The solid-state IR spectrum of complex 12.

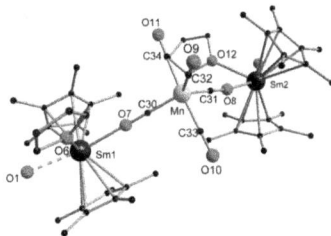

Figure 3.2.15: Cut-out of the polymeric structure of 12 in the solid state. H atoms and non-coordinating solvent molecules are omitted for clarity. Selected bond distances (Å) and angles [°]: Sm1-O1-2.463(3), Sm1-O6 2.518(3), Sm1-O7 2.449(3), Sm2-O12 2.555(3), Sm2-O8 2.458(3), Sm2-O2 2.474(3), Mn2-C30 1.762(4), Mn2-C31 1.772(4), Mn2-C32 1.831(5), Mn2 C33 1.834(5), Mn2-C34 1.845(5), O7-C30 1.184(5), O8-C31 1.181(5), O9-C32 1.149(6), O10-C33 1.144(6), O11-C34 1.136(6); O1-Sm1-O6 74.34(10), O6-Sm1-C7 91.00(13), C30-Mn2-C31 121.7(2), C30-Mn2-C32 119.2(2), C30-Mn2-C33 89.1(2), C30-Mn2-C34 88.82(19), C31-Mn2-C32 119.1(2), C31-Mn2-C33 91.0(2), C31-Mn2-C34 91.8(2), C32-Mn2-C33 90.9(2), C32-Mn2-C34 88.4(2), C33-Mn2-C34 177.1(2).

47

3.2.4 Theoretical calculations of complex 10 and 11

The theoretical calculations of complex 10 and 11 were performed and analysed by Dr. Ralf Köppe.

The theoretical calculations were performed with the RI-DFT/BP-86[156] method using def2-TZVP[157] basis sets for all atoms. For Re, as well as for Sm, effective core potentials containing 60 or 51 core electrons, respectively, were chosen. The population analyses based on occupation numbers were performed using 13 modified atoms orbitals (MAOs) on Re, 10 on Sm, 5 each on C and O and 1 MAO on H.[158] The program TURBOMOLE 7.3 was used to perform the quantum chemical calculations.[156a] The wave function for the Bader AIM analysis were carried out by wave function analyser multiwfn 3.6.[159] Over the years, it has been established that the bonding situation in many cases could be nicely explained by the calculation of the shared electron numbers (SEN) according to the population analysis of Ahlrichs and Heinzmann.[160] The SEN numbers give a reliable measure of the covalent bond strength, especially even in the case of multi-centre bonding contributions. In several complexes, formation of a covalent bond between two metals may be promoted by the presence of supporting ligands that are bridging the metal centres. This phenomenon is commonly referred as "supported metal-metal bonding". This effect might be seen in case of complex 10 and 11 where the Re-Re distances are influenced by the presence of the supporting bridging Re-C-Re. The reference value of SEN for an unsupported Re-Re single bond was calculated for a complex containing an unbridged Re-Re single bond, $[Re_2(CO)_{10}]$. The SEN (Re-Re) in the case of $[Re_2(CO)_{10}]$ is 1.297. However, a multi-centre bond strengthening corresponding to a four-centre contribution of the SEN (C-Re-Re-C) = 0.248 is also considered, which is about 19% of the total value of the SEN for the Re-Re bond. This procedure was tested and justified by Ahlrichs *et al.* for the molecule diborane.[158] After considering the multi-centre bond strengthening, the adjusted value for SEN of Re-Re bond would thus be 1.049. The anion $[Re_2Cl_9]^-$ was investigated as reference for a molecule with a triple-bridged formal Re-Re triple bond. The total calculated SEN (Re-Re) value is 1.915 in case of $[Re_2Cl_9]^-$ anion, including the contribution from the three-centre SEN (Re-Cl-Re) which is 0.273. Hence, the pure Re-Re "triple" bond after excluding the three-centre bonding effects caused by the three Re-Cl-Re bridges would result a value of $1.915 - 3 \times 0.273 = 1.096$.

This method was applied analogously to calculate the strength of the Re-Re covalent bond in complexes **10** and **11**. The following conclusions could be drawn: in complex **11**, the total SEN (Re-Re) value is 1.071 and the SEN value from the three-centre contribution (Re-Re-C) is 0.290. Therefore, the corrected SEN (Re-Re) value should be 0.491 after excluding the multi-centre contribution. In case of complex **10**, the total SEN (Re-Re) value is 1.394 and the three-centre contribution results in the SEN (Re-Re-C) value of 0.348, so that the corrected SEN (Re-Re) value is only 0.698, which corresponds to a rather weak single bond between Re-Re atoms. The outcome of the above calculations is concluded in the following points: (i) Re-Re bonding regardless of its origin evaluates in the following row (SEN values in parentheses): **11** (1.071) < [Re$_2$(CO)$_{10}$] (1.297) < **10** (1.394) < [Re$_2$Cl$_9$]$^-$ (1.915), and (ii) by considering the role of supporting CO ligands in the Re-Re contacts, we conclude that there is virtually no Re-Re bond in **11** and only a weak single two-centre Re-Re bond in **10**.

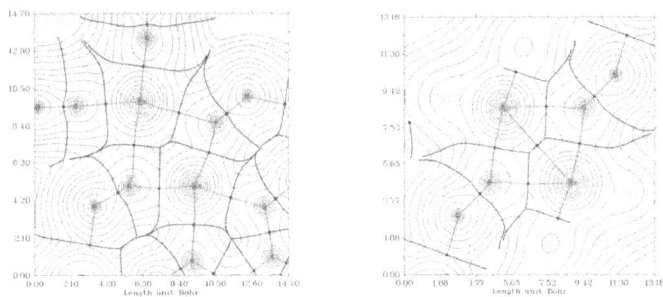

Figure 3.2.16: Contour plots of the electron densities of **11** (left) and **10** (right) each in the plane Re-Re-C obtained by AIM analysis of complex **11** and **10**. The connections of Re and C in each molecule and between Re and Re in **10** are depicted by the bond paths (orange) and the bond critical points (blue). Ring critical points are given in orange. The blue line represents the zero-flux surface that defines the sub space of each atom.

To further confirm the findings of the population analyses in an orienting manner, a different topological approach was performed by means of the AIM (atoms in molecules) method introduced by Bader.[161] The results of the contour plots of electron densities obtained by the AIM analysis are shown in Figure 3.2.16. Compound **10**, unlike **11**, reveals a bond path with a bond critical point between both rhenium atoms that undoubtedly confirms the presence of a Re-Re bond. The low value of the electron density ρ_{bcp} and the high value for the ellipticity ε (ρ_{bcp}

= 0.05 au, ε = 1.31) calculated for this bond critical point of Re-Re indicate a rather weak bond with a high "multiple bond contribution" presumably due to the interaction with the supporting π-type Re-Re-C multicentre bonds.

The bonding properties of the rhenacycle in compound **11** are also interesting to examine theoretically. The values for the SEN of the bonds C1-C2, C2-C3, C3-C4 of 1.514, 1.593, 1.518 correspond to strong single-bonds. The SEN values of about 1.4 or 2.28[158] are expected for pure single or double bonds, respectively. The SEN (Re-C) is calculated to be 0.95 which is of the same order of magnitude as for the terminal Re-CO bond (1.276). Overall, these results support the view of a double Fischer carbene complex as shown in **11a** (Scheme 3.2.6).

3.3 Reactivity of pentaphosphaferrocene with low-valent main group compounds

The $^{31}P\{^1H\}$ NMR simulations in this part of this thesis were performed by Dr. Thomas Simler.

3.3.1 Introduction

Phosphorous containing organo-heterocyclic compounds are very well known[162] and have found applications in material science,[163] pharmaceutical industries,[164] and in ligand design for coordination chemistry.[165] However, heterocyclic compounds comprising phosphorous and heavier group-14/13 elements are rather scarce.[166] Recently, the functionalization of white phosphorous with main group reagents for the direct access to the phosphorous heterocycles has attracted a wide-spread attention.[167] Indeed, low-valent main group compounds are highly susceptible to redox reactions and have been utilized to reduce white phosphorous and generate heterocyclic compounds.[168] For example, reactions of different carbenes with white phosphorous resulted in organo-phosphorous cages, chains, and rings.[169] Similarly, the reaction of white phosphorous with silylenes furnished sila-phospha cages and rings.[170] Also, monovalent aluminium complexes have been used for the activation of white phosphorous to obtain Al-P clusters and cages.[119]

[Cp*Fe(η^5-P$_5$)] has a very unique coordination chemistry due its penta-coordination ability owing to the C_5 symmetry of the *cyclo*-P$_5$ ring attached to the [Cp*Fe]$^+$ moiety.[72-74,76,79,171] Recent studies have revealed the involvement of the *cyclo*-P$_5$ ring during the reaction of [Cp*Fe(η^5-P$_5$)] with single electron reductants (divalent lanthanide complexes[82-84] and elemental potassium[81]; section 1.4.2) and main-group nucleophiles[86] (section 1.4.2). Although the reactivity of [Cp*Fe(η^5-P$_5$)] has been studied with reducing agents and nucleophiles, the substitution or insertion of main group heteroatom/s in the *cyclo*-P$_5$ ring has never been reported to date. Since low-valent group-13 and -14 compounds can insert in different types of P-P bonds,[119,167-168,170] it would be interesting to see whether a similar reactivity would also occur on [Cp*Fe(η^5-P$_5$)] and allow the substitution or insertion of other elements such as Si, Ge, and Al in the *cyclo*-P$_5$ ring. In this context, we examined the reactivity of [Cp*Fe(η^5-P$_5$)] with low-valent compounds from the groups 2, 13, and 14. In a similar way as in the previous chapter, the effects of different ligands and reducing agents with different reduction potentials have been investigated.

51

Group-14

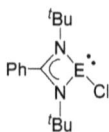

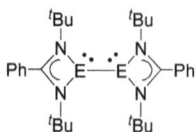

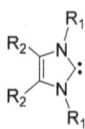

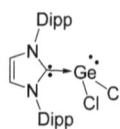

E = Si, Ge   E = Si, Ge R_1 = CH_3; R_2 = CH_3
                                                        or
                                               R_1 = Dipp; R_2 = H

Group-13

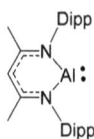

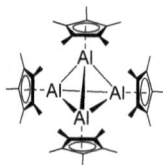

Group-2

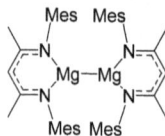

Chart 3.3.1: Different types of low-valent compounds used to react with [Cp*Fe(η^5-P_5)].

3.3.2 Reactivity of [Cp*Fe(η^5-P$_5$)] with *N*-heterocyclic carbenes

The activation of white phosphorous with different *N*-heterocyclic carbenes (NHCs) has enabled the isolation of phosphorous-rich organic cages, clusters, chains, and rings.[95,167] So far, the reactivity of [Cp*Fe(η^5-P$_5$)] with carbenes, which correspond to a first class of low-valent main group compounds, has not been studied. The reaction of 1,3-bis(2,6-diisopropylphenyl)imidazol-2-ylidene (IPr)[172] with [Cp*Fe(η^5-P$_5$)] was carried out at room temperature. However, the ^{1}H and ^{31}P{^{1}H} NMR spectra of the reaction mixture only showed resonances corresponding to the starting materials. The reaction mixture was then heated to 80 °C for 12 hours (Scheme 3.3.1) but, even at the elevated temperatures only a mixture of starting materials was detected in the ^{1}H NMR spectrum (Figure 3.3.1).

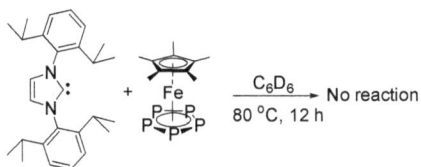

Scheme 3.3.1: The reaction between IPr and [Cp*Fe(η^5-P$_5$)].

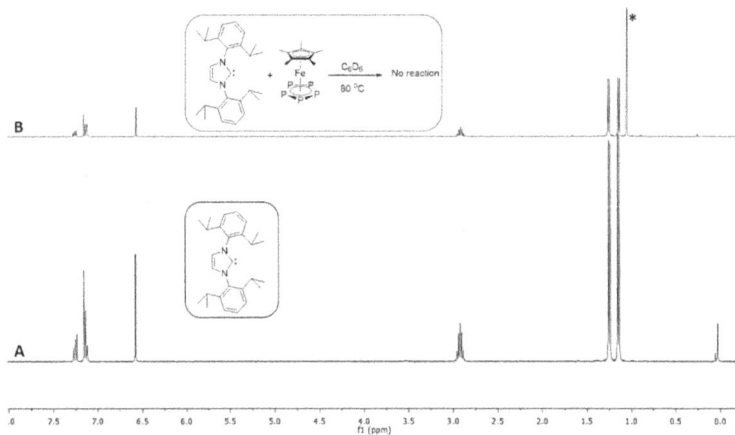

Figure 3.3.1: Stacked ^{1}H NMR spectra of **A** (only IPr) and **B** (IPr with [Cp*Fe(η^5-P$_5$)]). * = [Cp*Fe(η^5-P$_5$)].

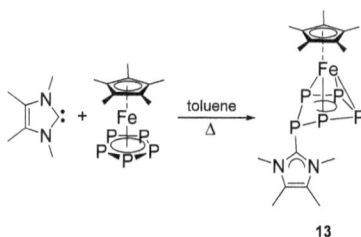

Scheme 3.3.2: Synthesis of complex **13**.

A possible reason for the unreactive nature of IPr towards [Cp*Fe(η^5-P$_5$)] may be the low nucleophilicity of the IPr carbene centre due to the steric bulk of the Dipp groups on the N atoms. This problem could be resolved by using a smaller and more nucleophilic carbene, namely 1,3,4,5-tetramethylimidazol-2-ylidene (ITMe).[173] Reaction of ITMe with [Cp*Fe(η^5-P$_5$)] in toluene (Scheme 3.3.2) led to a clear green-coloured solution, which subsequently furnished green-coloured crystals of complex **13** after standing at room temperature for a few days. Analytically pure complex **13** could be isolated in 92% yield. The molecular structure of [ITMe{(η^4-P$_5$)FeCp*}] (**13**) was determined by X-ray diffraction studies (Figure 3.3.2). Complex **13** crystallizes in the monoclinic space group $P2_1/n$ with one molecule in the asymmetric unit cell. The solid-state structure of **13** shows an envelope-shaped cyclo-P$_5$ ring η^4-coordinated to the [Cp*Fe]$^+$ moiety and formation of a P-C bond between the carbene and one P atom of the cyclo-P$_5$ ring. The Fe-P bond lengths (2.291(2)-2.330(2) Å) in complex **13** are shorter than those in [Cp*Fe(η^5-P$_5$)] (2.3608(11)-2.3726(11) Å),[79] but similar to the Fe-P bond distances in previously reported complexes containing an envelope-shaped cyclo-P$_5$ ring, for example [{K$_2$}(Cp*Fe(η^4-P$_5$))] (average 2.291 Å).[81] The relatively short P-P bond lengths in the cyclo-P$_5$ ring (2.105(5)-2.162(3) Å) suggest a partial double bond character.[174] The P-C bond length (1.862(6) Å) is in the typical range for single bonds between P and C atoms.[86] The ^{1}H NMR spectrum of **13** is consistent with the solid-state structure, showing three singlets at δ 0.77, 1.84, and 2.86 ppm. The ^{31}P{^{1}H} NMR spectrum of **13** shows two broad resonance at δ -47.5 and 37.8 ppm at room temperature, suggesting a fluxional behaviour of the cyclo-P$_5$ ring (Figure 3.3.3). Measurement of the ^{31}P{^{1}H} NMR spectrum at lower temperature (233 K) shows a well-resolved AMM'XX' spin system with multiplets at δ -51.0 (P$_{XX'}$), 33.1 (P$_{MM'}$), and 39.2 (P$_A$) ppm. The multiplets were assigned by higher order P-P coupling and the coupling constants were obtained from an iterative fitting of the $^{31}P\{^1H\}$ NMR spectrum (Figures 3.3.4 and Table 3.3.1).

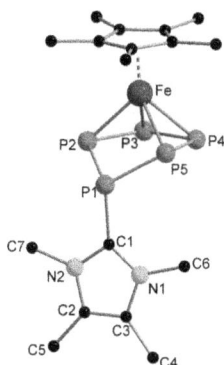

Figure 3.3.2: The molecular structure of **13** in solid state. Hydrogen atoms are omitted for clarity. Selected bond distances (Å) and angles [°]: Fe-P2 2.305(2), Fe-P3 2.330(2), Fe4 2.327(3), Fe-P5 2.291(2), Fe-C8 2.099(7), Fe-C9 2.087(7), Fe-C10 2.070(7), Fe-C11 2.065(7), Fe-C12 2.097(8), P1-P2 2.162(3), P1-P5 2.135(3), P1-C1 1.862(6), P2-P3 2.146(4), P3-P4 2.105(5), P4-P5 2.151(4), N1-C1 1.340(10), N1-C3 1.378(9), N1-C6 1.472(11), N2-C1 1.341(9), N2-C2 1.400(9), N2-C7 1.484(10), C2-C3 1.369(12), C2-C5 1.468(11), C3-C4 1.485(11); P5-P1-P2 94.54(12), C1-P1-P2 112.8(2), C1-P1-P5 115.3(3), P3-P2-P1 108.0(2), P4-P3-P2 103.95(14), P3-P4-P5 104.4(2), P1-P5-P4 107.8(2), N1-C1-N2 105.6(5).

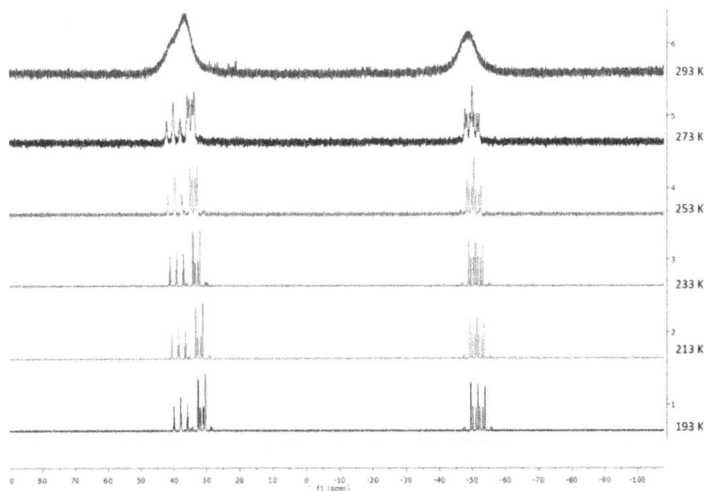

Figure 3.3.3: Variable temperature $^{31}P\{^1H\}$ NMR (162 MHz, thf-d_8) spectra of complex **13**.

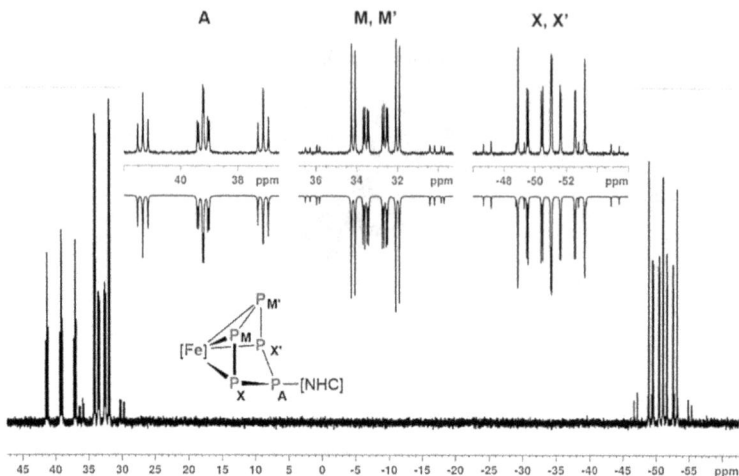

Figure 3.3.4: ^{31}P$\{^1$H$\}$ NMR spectrum at 233 K of compound **13** with nuclei assigned to an AMM'XX' spin system; insets: extended signals (upwards) and simulations (downwards); δ (P$_A$) = 39.2 ppm, δ (P$_{MM'}$) = 33.1 ppm, δ (P$_{XX'}$) = -51.0 ppm, $^1J_{MM'}$ = 419.3 Hz, $^1J_{MX}$ = $^1J_{M'X'}$ = 368.1 Hz, $^1J_{AX}$ = $^1J_{AX'}$ = -342.6 Hz, $^2J_{XX'}$ = 47.4 Hz, $^2J_{MX'}$ = $^2J_{M'X}$ = -16.2 Hz, $^2J_{AM}$ = $^2J_{AM'}$ = 29.7 Hz, [Fe] = Cp*Fe, [NHC] = ITMe.

Table 3.3.1: Chemical shifts, couplings constants and linewidths from the iterative fit of the AMM'XX' spin system of **13**.

Parameters	Iteration values	Parameters	Iteration values
δ (P$_A$)	39.19 ppm	$^1J_{MM'}$	419.3 Hz
δ (P$_M$) = δ (P$_{M'}$)	33.07 ppm	$^1J_{MX}$ = $^1J_{M'X'}$	368.1 Hz
δ (P$_X$) = δ (P$_{X'}$)	-51.00 ppm	$^1J_{AX}$ = $^1J_{AX'}$	-342.6 Hz
$\omega_{1/2}$ (A)	3.5 Hz	$^2J_{XX'}$	47.4 Hz
$\omega_{1/2}$ (M) = $\omega_{1/2}$ (M')	4.1 Hz	$^2J_{MX'}$ = $^2J_{M'X}$	-16.2 Hz
$\omega_{1/2}$ (X) = $\omega_{1/2}$ (X')	3.6 Hz	$^2J_{AM}$ = $^2J_{AM'}$	29.7 Hz

3.3.3 Reactivity of [Cp*Fe(η^5-P$_5$)] with *N*-heterocyclic silylenes

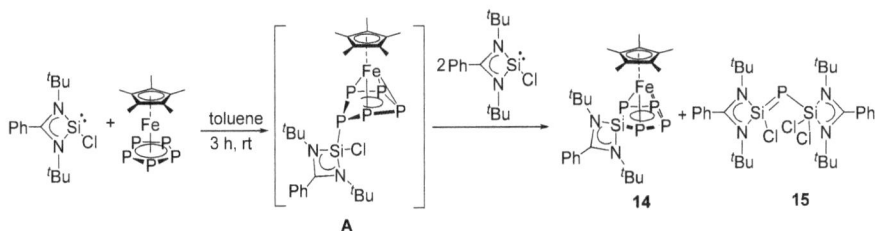

Scheme 3.3.3: Synthesis of compound **14** and **15** *via* the possible intermediate **A**.

Similarly to NHCs, NHSis have been employed for metal-free activation of white phosphorous. By using different NHSis with white phosphorus different Si-P compounds have been isolated.[175] For example, the Si-P heterocyclic compound [L$_2$Si$_2$P$_2$] (**L** = PhC(N^{t}Bu)$_2$), consisting in an antiaromatic four-membered Si-P ring, was isolated by reaction of LSiCl with white phosphorous.[176] The formation of [L$_2$Si$_2$P$_2$] could be explained by the initial insertion of a silylene moiety in one P-P bond of tetrahedral P$_4$. In a second step, another silylene abstracts the chloride of the intermediate, which results in the formation of the four-membered ring. Whether a similar insertion of a silicon moiety can be achieved in a non-strained polyphosphide system is thus interesting to investigate. To pursue this idea, the reactivity of [Cp*Fe(η^5-P$_5$)] was examined towards LSiCl. The reaction of LSiCl[177] with [Cp*Fe(η^5-P$_5$)] in a 1:1 molar ratio was carried out in toluene at room temperature. The ^{1}H and ^{31}P{^{1}H} NMR spectra of the reaction mixture showed full consumption of LSiCl and the presence of new signals. A significant amount of unreacted [Cp*Fe(η^5-P$_5$)] was also detected, which suggests that more than one equivalent of LSiCl is required for the full consumption of one equivalent of [Cp*Fe(η^5-P$_5$)]. Full conversion could be achieved using a 3:1 molar ratio between LSiCl and [Cp*Fe(η^5-P$_5$)], respectively (Scheme 3.3.3), and the reaction led to the isolation of two major products, [(η^4-P$_4$SiL)FeCp*] (**14**) and LSi(Cl)=P-SiL(Cl)$_2$ (**15**). The separation of both species was enabled by their slightly different solubilities in hydrocarbon solvents. The reaction mixture was first extracted with hexane and filtered, leading to compound **15** in a 32% crystalline yield by storing the hexane solution at -30 °C. The residue of the hexane extraction was further extracted with toluene, filtered and stored at -30 °C, resulting in the formation of brown-coloured crystals of complex **14** in 40% yield. The solid-state structure of complex **14** was determined by X-ray diffraction studies. Complex **14** crystallizes in the monoclinic space group $P2_1/c$ with one molecule in the asymmetric unit cell. The molecular structure shows the formation of a five-membered [*cyclo*-LSiP$_4$]$^-$ ring η^4-coordinated to the [Cp*Fe]$^+$ moiety (Figure 3.3.5). The Fe-P1 (2.397(10) Å) and Fe-P4 (2.379 (10) Å) bond distances are significantly longer than the Fe-P2 (2.322(9) Å) and Fe-P3 (2.327 (9) Å) analogues, possibly due to the coordination of P1 and P4 to the Si atom. The Si atom is in a distorted tetrahedral environment, coordinated by the two N atoms of the amidinate ligand and two P atoms of the five-membered ring. The Si-P1 (2.171(13) Å) and Si-P4 (2.173(12) Å) bond distances are almost identical and lie in-between single (2.06-2.09 Å) and double bonds (2.24-2.27 Å).[174,176,178] Besides, the P-P bond lengths (P1-P2 2.139(13), P2-P3 2.159(15), and P3-P4 2.144(13) Å) show a partial double-bond character, as observed in previously reported [K$_2$·{Cp*Fe(η^4-P$_5$)}] (2.133(1)-2.153(1) Å).[81] As a result, the [*cyclo*-LSiP$_4$]$^-$ can be considered as a 6π-electron aromatic five-membered ring. However, in comparison to [*cyclo*-P$_5$]$^-$, the five-membered [*cyclo*-LSiP$_4$]$^-$ ring is not planar, which can be accredited to the steric repulsion between one ^{t}Bu-group of the amidinate ligand and the methyl groups on the [Cp*Fe]$^+$ moiety. The dihedral angle between the planes defined by P1-P2-P3-P4 and P1-Si-P4 is 53.7°.

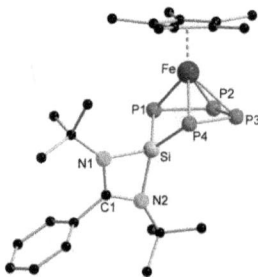

Figure 3.3.5: Molecular structure of **14** in the solid state. H atoms are omitted for clarity. Selected bond distances (Å) and angles [°]: Fe-P1 2.3795(10), Fe-P2 2.3222(9), Fe-P3 2.3271(9), Fe-P4 2.3973(10), P1-P2 2.1391(13), P2-P3 2.160(2), P3-P4 2.1444(13), Fe-C16 2.107(3), Fe-C17 2.122(3), Fe-C18 2.094(3), Fe-C19 2.050(3), Fe-C20 2.068(3), P1-Si 2.1707(13), P4-Si 2.1729(12), Si-N1 1.832(2), Si-N2 1.849(3), N1-C1 1.333(4), N2-C1 1.335(4); P1-Si-P4 103.78(5), N1-Si-P1 112.25(10), N1-Si-P4 117.47(10), N1-Si-N2 71.35(11), N2-Si-P1 120.75(11), N2-Si-P4 127.61(10), P1-P2-P3 107.22(5), P4-P3-P2 106.93(5), N1-C1-N2 107.1(2).

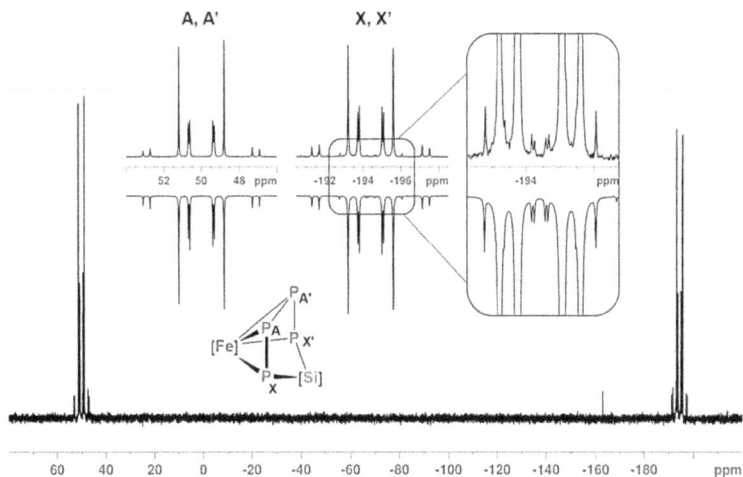

Figure 3.3.6: ^{31}P{^{1}H} NMR spectrum (162.0 MHz, 298 K, C$_6$D$_6$) of compound **14** with nuclei assigned to an AA′XX′ spin system; insets: extended signals (upwards) and simulations (downwards); δ (P$_{AA'}$) = 50.0 ppm, δ (P$_{XX'}$) = -194.4 ppm, $^1J_{AA'}$ = 364.4 Hz, $^1J_{AX}$ = $^1J_{A'X'}$ = 415.9 Hz, $^2J_{AX'}$ = $^2J_{A'X}$ = -27.7 Hz, $^2J_{XX'}$ = 37.2 Hz, $^1J_{SiX}$ = $^1J_{SiX'}$ = 145.5 Hz, $^2J_{SiA}$ = $^2J_{SiA'}$ = 3.1 Hz, [Fe] = Cp*Fe, [Si] = PhC(N^{t}Bu)$_2$Si.

Table 3.3.2: Chemical shifts, couplings constants and linewidths from the iterative fit of the AA′XX′ spin system of **14**.

Parameters	Iteration values	Parameters	Iteration values
δ (P$_A$) = δ (P$_{A'}$)	50.00 ppm	$^1J_{XA}$ = $^1J_{X'A'}$	415.9 Hz
δ (P$_X$) = δ (P$_{X'}$)	-194.41 ppm	$^2J_{XX'}$	37.2 Hz
$\omega_{1/2}$ (A) = $\omega_{1/2}$ (A′)	3.9 Hz	$^2J_{XA'}$ = $^2J_{X'A}$	-27.7 Hz
$\omega_{1/2}$ (X) = $\omega_{1/2}$ (X′)	5.2 Hz	$^1J_{SiX}$ = $^1J_{SiX'}$	145.5 Hz
$^1J_{AA'}$	364.4 Hz	$^2J_{SiA}$ = $^2J_{SiA'}$	3.1 Hz

The ^{1}H NMR spectrum of complex **14** shows three singlets at δ 0.67, 1.27, and 1.85 ppm in the aliphatic region. The resonance at δ 1.85 ppm corresponds to the methyl protons of the [Cp*Fe]$^+$ group, which is downfield shifted relative to [Cp*Fe(η^5-P$_5$)] (δ 1.08 ppm). The signals at δ 0.67 and 1.27 ppm are assigned to the two ^{t}Bu groups in different chemical environments, indicating a rigid conformation of the amidinate ligand on the tetrahedrally coordinated Si atom. The ^{31}P{^{1}H} NMR spectrum of complex **14** shows an AA'XX' spin system with two sets of multiplets at δ -194.4 and 50 ppm (Figure 3.3.6). Such an AA'XX' spin system is typical for an open-chain P$_4$ unit, in agreement with the solid-state structure of **14**.[179] The multiplet at δ -194.4 ppm can be unambiguously assigned to the P1 and P4 atoms (P$_{XX'}$) due to the presence of well-defined Si satellite peaks with a $^1J_{Si\text{-}P}$ coupling 145.5 Hz (Table 3.3.2). Accordingly, the ^{29}Si{^{1}H} NMR spectrum reveals a triplet at δ 42.3 ppm ($^1J_{Si\text{-}P}$ = 145.4 Hz). The different coupling constants were determined by an iterative fitting of the ^{31}P{^{1}H} spectrum. The $^1J_{AX}$ coupling constant (415.9 Hz) is about 70 Hz larger in magnitude than the $^1J_{AA'}$ coupling constant (364.4 Hz), a trend that has already been observed by Wolf and co-workers in a Ga-Co polyphosphide complex featuring a bridging P$_4$ chain unit.[179e] The formation of complex **14** results from the formal insertion of a [LSi]$^+$ moiety in the *cyclo*-P$_5$ ring of [Cp*Fe(η^5-P$_5$)], accompanied by the elimination of one P atom from the *cyclo*-P$_5$ ring and of one Cl atom from LSiCl. The eliminated P and Cl atoms are scavenged by two other molecules of LSiCl, forming a base-stabilized phosphasilene (**15**), which identity was unambiguously established by X-ray diffraction studies (Figure 3.3.7). Compound **15** crystallizes in the triclinic space group $P\bar{1}$ with one molecule in the asymmetric unit cell. The Si2 atom is in a distorted tetrahedral environment, coordinated by the two N atoms of an amidinate ligand, one Cl atom, and one P atom. In contrast, the Si1 atom lies in a five-coordinate environment, coordinated by the P atom, one chelating amidinate and two Cl ligands. The Si2-P bond length (2.109(8) Å) is in the range of reported Si=P bonds.[176,178] Surprisingly, the Si1-P bond distance (2.192(8) Å) is in between the distances for Si-P single and double bonds, suggesting a partial double-bond character. Such a shortening may be the result of the donation of the P lone pair to Si1. This hypothesis is further supported by the elongation of the Si1-N2 bond (Si1-N2 1.98(2) *vs* Si1-N1 1.81(2) Å) and the shortening of the N2-C1 bond (N2-C1 1.305(2) *vs* N1-C1 1.361(2) Å).

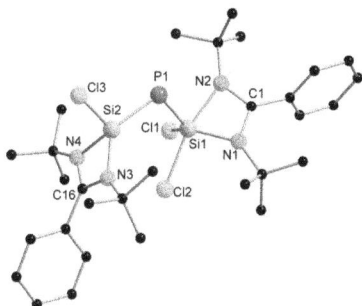

Figure 3.3.7: Molecular structure of **15** in the solid state. H atoms are omitted for clarity. Selected bond distances (Å) and angles [°]: P1-Si2 2.1093(8), P1-Si1 2.1925(8), Si1-N1 1.809(2), Si1-N2 1.9823(2), Si2-N3 1.8014(2), Si2-N4 1.8054(2), Cl1-Si1 2.1188(8), Cl2-Si1 2.2173(7), Cl3-Si2 2.0821(8), N1-C1 1.361(2), N2-C1 1.305(2), N3-C16 1.326(3), N4-C16 1.325(3); Si2-P1-Si1 105.61(3), Cl1-Si1-Cl2 89.58(3), Cl1-Si1-P1 123.70(3), P1-Si1-Cl2 102.58(3), N1-Si1-Cl1 109.57(6), N1-Si1-Cl2 97.14(6), N1-Si1-P1 122.64(6), N1-Si1-N2 68.81(7), N2-Si1-Cl1 88.09(5), N2-Si1-Cl2 163.96(5), N2-Si1-P1 91.88(5), Cl3-Si2-P1 106.87(3), N3-Si2-Cl3 102.65(6), N3-Si2-P1 133.35(6), N3-Si2-N4 71.98(8), N4-Si2-Cl3 104.67(6), N4-Si2-P1 131.11(6), N2-C1-N1 107.3(2), N4-C16-N3 106.20(2).

It should be noted that only one singlet at δ 1.47 ppm for the protons of ^{t}Bu groups was detected in the ^{1}H NMR spectrum of **15**, suggesting the equivalence of both amidinate ligands on the NMR time-scale. The ^{31}P{^{1}H} NMR spectrum showed a singlet at δ -182.7 ppm featuring two Si satellites with $^1J_{Si-P}$ = 84.4 Hz. In solution, complex **15** slowly decomposes to two known species, LSiCl$_3$[177a] and the recently reported L$_2$Si$_2$P$_2$ (Scheme 3.3.4).[176,178b,178c] Owing to the low signal-to-noise ratio resulting from the coupling of the Si atoms with the P atom, and decomposition of **15** in solution, the signals for the Si atoms in the ^{29}Si{^{1}H} NMR spectrum could not be detected even using extended acquisition times.

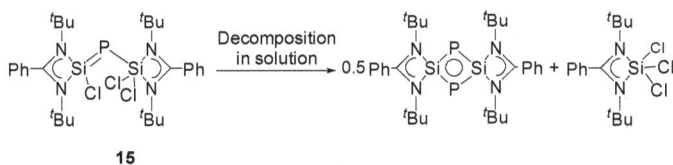

Scheme 3.3.4: Decomposition of compound **15** in solution at room temperature.

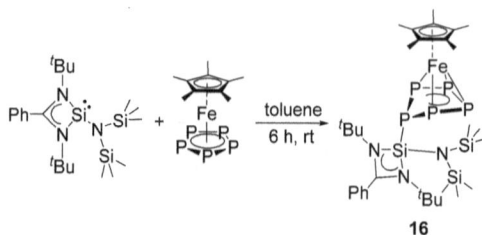

Scheme 3.3.5: Synthesis of complex 16.

The isolation of **14** and **15** from the reaction between LSiCl and [Cp*Fe(η^5-P$_5$)] suggested that **A** could be a reaction intermediate (Scheme 3.3.3). We reasoned that by using a derivative of LSiCl bearing bulky substituents and a lesser-labile group than Cl, stabilization of the intermediate might be possible. For that purpose, the reaction between [LSi(N(SiMe$_3$)$_2$)][180] and [Cp*Fe(η^5-P$_5$)] was carried out and resulted in the formation of [{LSi(N(SiMe$_3$)$_2$)}{(η^4-P$_5$)FeCp*}] (**16**) (Scheme 3.3.5) in 79% yield. Complex **16** crystallizes in the triclinic space group $P\bar{1}$ with one molecule in the asymmetric unit cell. Its molecular structure shows the transformation of the *cyclo*-P$_5$ ring from planar to envelope-shaped (Figure 3.3.8), which renders complex **16** structurally analogous to the postulated intermediate **A**. The Si atom in [LSi(N(SiMe$_3$)$_2$)] is oxidized from the +2 to +4 oxidation state, resulting from the transfer of 2e$^-$ from the silylene to [Cp*Fe(η^5-P$_5$)]. The two generated fragments, [LSi(N(SiMe$_3$)$_2$)]$^{2+}$ and [Cp*FeP$_5$]$^{2-}$, combine together *via* formation of a Si-P bond. The 2e$^-$ reduction of [Cp*Fe(η^5-P$_5$)] by [LSi(N(SiMe$_3$)$_2$)] is consistent with the formation of an envelope-shaped *cyclo*-P$_5$ ring.[81-82] The Si1 atom is pentacoordinated, surrounded by the two N atoms of the amidinate ligand, one N atom of the [N(SiMe$_3$)$_2$] group, and one P atom. The Si1-P1 bond length (2.272(5) Å) is similar to that in single Si-P bonds. The [Cp*Fe]$^+$ fragment is η^4-coordinated to the *cyclo*-P$_5$ ring. The average Fe-P bond distance (2.306 Å) is similar to the Fe-P bond distances in other di-reduced systems such as [K$_2$·{Cp*Fe(η^4-P$_5$)}] (average 2.291 Å).[81] The P-P bond distances in complex **16** are ranging from 2.125(9) to 2.172(6) Å, suggesting a partial double-bond character.[81,174,178a,178c] Interestingly, at room temperature, the ^{31}P{^{1}H} NMR spectrum of **16** shows a broad signal at δ 35.3 ppm, indicating a fluxional behaviour of the *cyclo*-P$_5$ ring. This dynamic behaviour was confirmed by variable temperature NMR studies, as more resolved signals could be obtained when the spectrum was recorded at lower temperatures

(Figure 3.3.9). At 193 K, a well-resolved spectrum with signals at δ -29.9 ($P_{XX'}$), 28.4 (P_M), and 30.7 ($P_{AA'}$) ppm could be observed and is consistent with an AA'MXX' spin system typical for envelope-shaped $cyclo$-P_5 systems (Figure 3.3.10). The different coupling constants were extracted by the iterative simulation of the $^{31}P\{^1H\}$ NMR spectrum (Table 3.3.3). The out-of-plane phosphorus coordinated to the silicon moiety is detected as a pseudo triplet of triplets at δ 28.4 ppm with a $^1J_{MX}$ coupling constant (-360.1 Hz) comparable to that reported for related systems. The $^{29}Si\{^1H\}$ NMR spectrum exhibits three resonances at δ = -34.9, 6.0, and 10.8 ppm. The higher frequency signals can be assigned to the two trimethylsilyl groups and the resonance at δ -34.93 ppm corresponds to the Si1 atom. No $^1J_{Si-P}$ coupling could be seen at room temperature, possibly due to the fluxional behaviour of the $cyclo$-P_5 ring.

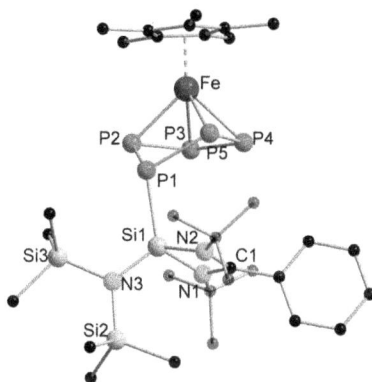

Figure 3.3.8: Molecular structure of **16** in the solid state. H atoms and non-coordinating solvent molecules are omitted for clarity. Selected bond distances (Å) and angles [°]: Fe-P2 2.2913(5), Fe-P3 2.3221(6), Fe-P4 2.3131(6), Fe-P5 2.2981(5), Fe-C22 2.0598(2), Fe-C23 2.0775(16), Fe-C24 2.1031(2), Fe-C26 2.1021(2), Fe-C27 2.0702(2), P1-P2 2.1725(6), P1-P5 2.1654(6), P2-P3 2.1483(8), P3-P4 2.1252(10), P4-P5 2.1533(9), P1-Si1 2.2722(5), Si1-N1 1.8127(13), Si1-N2 1.8166(13), Si1-N3 1.7062(13), Si3-N3 1.7831(13), N1-C1 1.3419(2), N2-C1 1.3446(2); P2-P1-Si1 120.54(2), P5-P1-Si1 110.71(2), P5-P1-P2 91.82(2), P3-P2-P1 108.91(3), P4-P3-P2 102.99(3), P3-P4-P5 103.64(3), P4-P5-P1 108.55(3), N1-Si1-N2 72.52(6), N1-C1-N2 106.07(12).

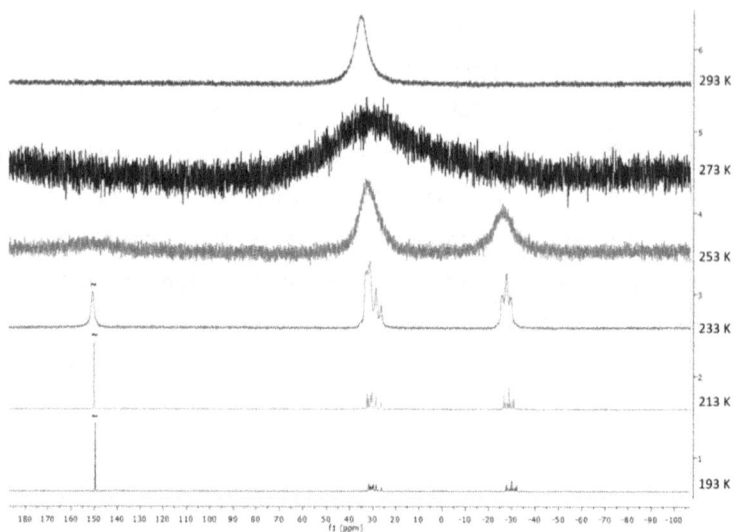

Figure 3.3.9: Variable temperature ^{31}P{^{1}H} NMR (162 MHz, toluene-d_8) spectra of complex **16**. $\sim$ = [Cp*Fe(η^5-P$_5$)].

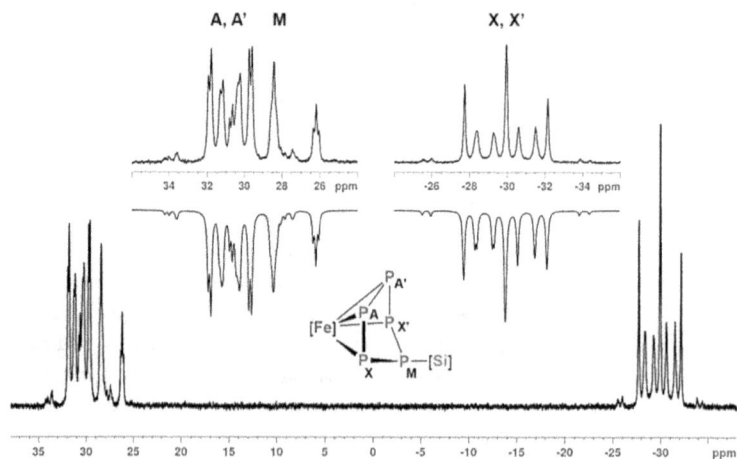

Figure 3.3.10: ^{31}P{^{1}H} NMR spectrum (162 MHz, 193 K, toluene-d8) of compound **16** with nuclei assigned to an AA′MXX′ spin system; insets: extended signals (upwards) and simulations (downwards); δ(P$_{AA'}$) = 30.7 ppm, δ(P$_M$) = 28.4 ppm, δ(P$_{XX'}$) = −29.9 ppm, $^1J_{AA'}$ = 424.4 Hz, $^1J_{AX}$ = $^1J_{A'X'}$ = −369.4 Hz, $^1J_{MX}$ = $^1J_{MX'}$ = −360.1 Hz, $^2J_{XX'}$ = 47.0 Hz, $^2J_{AX'}$ = $^2J_{A'X}$ = 14.4 Hz, $^2J_{AM}$ = $^2J_{A'M}$ = 24.9 Hz, [Fe] = Cp*Fe, [Si] = PhC(N^{t}Bu)$_2$SiN(SiMe$_3$)$_2$.

64

Table 3.3.3: Chemical shifts, couplings constants and linewidths from the iterative fit of the AA′MXX′ spin system of **16** at 193 K.

Parameters	Iteration values	Parameters	Iteration values
$\delta(P_A) = \delta(P_{A'})$	30.74 ppm	$^1J_{AA'}$	424.4 Hz
$\delta(P_M)$	28.39 ppm	$^1J_{AX} = {}^1J_{A'X'}$	-369.4 Hz
$\delta(P_X) = \delta(P_{X'})$	-29.90 ppm	$^1J_{MX} = {}^1J_{MX'}$	-360.1 Hz
$\omega_{1/2}(A) = \omega_{1/2}(A')$	19.2 Hz	$^2J_{XX'}$	47.0 Hz
$\omega_{1/2}(M)$	17.3 Hz	$^2J_{AX'} = {}^2J_{A'X}$	14.4 Hz
$\omega_{1/2}(X) = \omega_{1/2}(X')$	18.5 Hz	$^2J_{AM} = {}^2J_{A'M}$	24.9 Hz

3.3.4 Reactivity of [Cp*Fe(η^5-P$_5$)] with chloro-germylenes

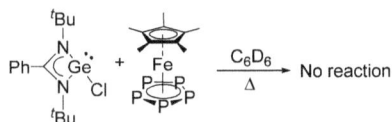

Scheme 3.3.6: Reaction of LGeCl with [Cp*Fe(η^5-P$_5$)].

As interesting results were observed in the investigation of the reactivity of [Cp*Fe(η^5-P$_5$)] with carbenes and silylenes, we moved onto their heavier analogues, germylenes, and LGeCl[181] was selected as a precursor. The reaction between LGeCl and [Cp*Fe(η^5-P$_5$)] was first attempted (Scheme 3.3.6). However, no reaction was observed even upon heating the reaction mixture at 80 °C for 24 hours, as evidenced by analysis of the ^{1}H and ^{31}P{^{1}H} NMR spectra of the reaction

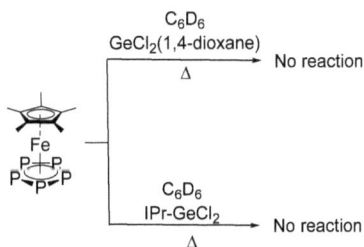

Scheme 3.3.7: Reaction of GeCl$_2$(1,4-dioxane) and [IPr-GeCl$_2$] with [Cp*Fe(η^5-P$_5$)].

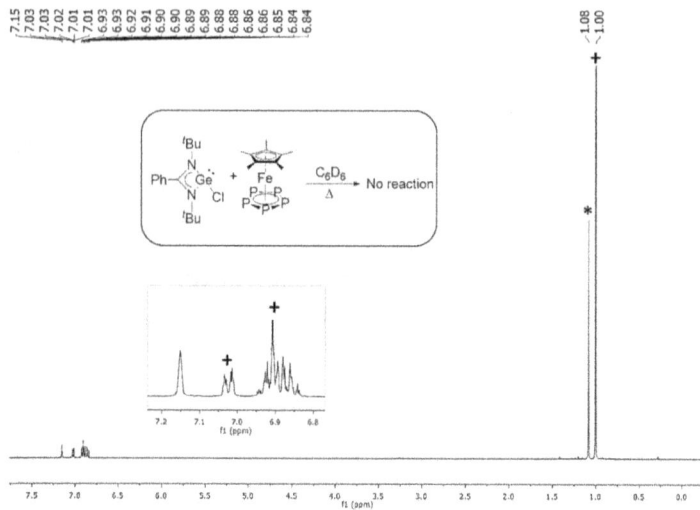

Figure 3.3.11: ^{1}H NMR spectrum of the reaction mixture between LGeCl (+) and [Cp*Fe(η^5-P$_5$)] (*).

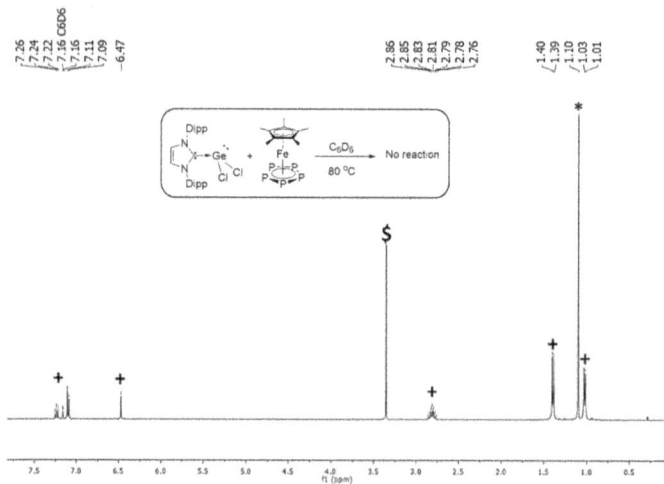

Figure 3.3.12: ^{1}H NMR spectrum of the reaction mixture between IPr-GeCl$_2$ (+) and [Cp*Fe(η^5-P$_5$)] (*). $ = 1,4-dioxane.

mixture (Figure 3.3.11). The unreactive nature of LGeCl towards [Cp*Fe(η^5-P$_5$)] can be due to the lower tendency of LGeCl to get oxidized compared to LSiCl and to the energetic stabilization of the lone pair in heavier group-14 elements (the inert pair effect). Very recently, Roesky et al. have shown a contrast in the reactivity of As-N bonds towards LGeCl and GeCl$_2$(1,4-dioxane). While the lithium salts of arsinoamides simply react with LGeCl through salt-metathesis reactions, As-N bond activation occurs when using GeCl$_2$(1,4-dioxane).[182] Inspired by this reactivity, the reaction of GeCl$_2$(1,4-dioxane) with [Cp*Fe(η^5-P$_5$)] was considered. Unfortunately, no reaction was observed even after heating the reaction mixture for several hours at 80 °C. To increase the reactivity of the germylene lone pair, the NHC-coordinated GeCl$_2$ complex [IPr-GeCl$_2$][183] was used in the reaction with [Cp*Fe(η^5-P$_5$)] (Scheme 3.3.7). However, in this case also, ^{1}H and ^{31}P{^{1}H} NMR analyses of the reaction mixture revealed no sign of reaction (Figure 3.3.12).

3.3.5 Reactivity of [Cp*Fe(η^5-P$_5$)] with di-silylene and di-germylene

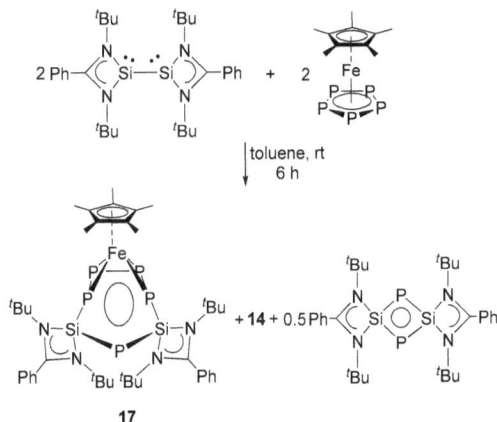

Scheme 3.3.8: Synthesis of complex 17.

After investigating the reactivity of [Cp*Fe(η^5-P$_5$)] with divalent carbenes, silylenes and germylenes, we were motivated to examine whether similar results would be obtained using formally monovalent tetrylenes. Di-silylene [LSi-SiL][184] and di-germylene [LGe-GeL][181] have their Si and Ge atoms in the formal +1 oxidation state and do not feature any leaving group, which makes them very interesting candidates for reactivity studies with [Cp*Fe(η^5-P$_5$)]. The reaction of [LSi-SiL] with one equivalent of [Cp*Fe(η^5-P$_5$)] in toluene at room temperature (Scheme 3.3.8) led to a colour change of the reaction mixture from green to reddish brown. Analysis of the ^{31}P{^{1}H} NMR spectrum of the crude reaction mixture revealed the formation of 14, L$_2$Si$_2$P$_2$, and signals corresponding to new species (Figure 3.3.13). After extraction with hexane and filtration, yellow-coloured crystals were isolated in 27% yield upon storing the hexane solution at -30 °C. X-ray analysis revealed the formation of [{(η^4-P$_5$(SiL)$_2$}FeCp*] (17) (Figure 3.3.15). The molecular structure of complex 17 showed the formation of a Si-P seven-membered ring in the coordination sphere of [Cp*Fe]$^+$. Complex 17 results from the insertion of two [LSi]$^+$ moieties into two adjacent P-P bonds of the cyclo-P$_5$ ring. After insertion, the Si atoms are formally oxidized from the +1 to +4 oxidation state. Therefore, complex 17 can be considered as a six-electron reduction product of [Cp*Fe(η^5-P$_5$)]. The ^{1}H NMR spectrum of complex 17 shows signals at δ 1.46 and 1.49 ppm for the ^{t}Bu groups, indicating no rotation of the amidinate ligand on the four-coordinated silicon. The protons of Cp* ligand in 17 are more downfield shifted as compared to those in 14 (δ = 1.96 vs 1.85 ppm, respectively). The ^{31}P{^{1}H} NMR spectrum of 17 shows an AA'MM'X spin system which was successfully simulated using an iterative fitting procedure (Figure 3.3.14), revealing two sets of multiplets for the four P atoms coordinated to the [Cp*Fe]$^+$ moiety at δ -47.6 (P$_{MM'}$) and 80.8 (P$_{AA'}$) ppm. The resonance of the P atom located between the two Si atoms (P$_X$) in the ring is upfield shifted to δ -163.5 ppm ($^1J_{Si-P}$ = 118.3 Hz). The multiplet at δ -47.6 ppm is unambiguously assigned to P2 and P5, on the basis of well-defined Si satellite signals ($^1J_{Si-P}$ = 169.3 Hz). In accordance with the ^{31}P{^{1}H} NMR spectrum, a doublet of doublet at δ 31.9 ppm with the same $^1J_{Si-P}$ coupling constants was detected in the ^{29}Si{^{1}H} NMR spectrum. To the best of our knowledge, 6e$^-$ reduction of [Cp*Fe(η^5-P$_5$)] has never been observed before. Complex 17 also represents the first example of a seven-membered Si-P ring η^4-coordinated to a [Cp*Fe]$^+$ moiety. The Si1-P2 (2.2151(8) Å) and Si2-P5 (2.2126(9) Å) bond lengths are slightly shorter than Si-P single bonds (2.24-2.27 Å).[174,178b] In contrast, the Si1-P1 (2.1419(8) Å) and Si2-P1 (2.1407(9) Å) distances are in-between single (2.24-2.27 Å) and double bonds (2.06-2.09 Å).[83,174,176,178]

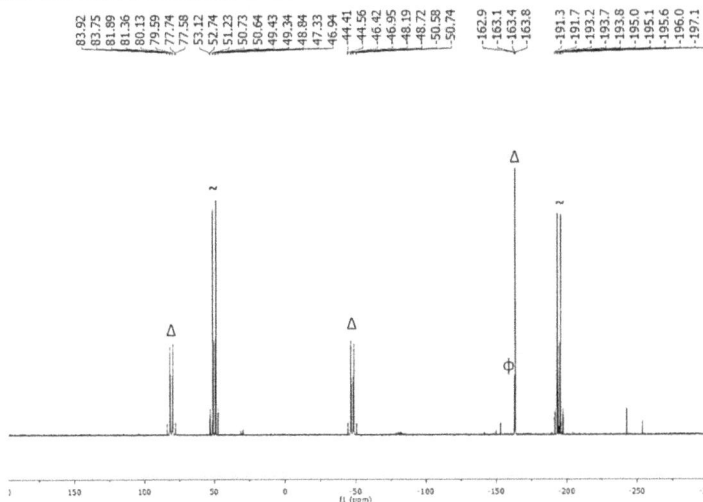

Figure 3.3.13: $^{31}P\{^1H\}$ NMR (162 MHz, 298 K, C_6D_6) spectrum of the crude mixture after reaction between [LSi-SiL] and [Cp*Fe(η^5-P$_5$)], $\sim$ = **14**, Δ = **17**, and ϕ = $L_2Si_2P_2$.

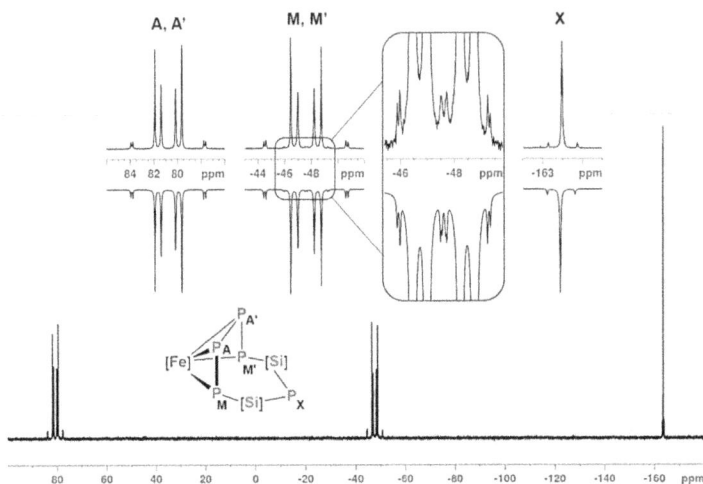

Figure 3.3.14: $^{31}P\{^1H\}$ NMR spectrum at 298 K of compound **17** with nuclei assigned to an AA'MM'X spin system; insets: extended signals (upwards) and simulations (downwards); δ (P$_{AA'}$) = 80.8 ppm, δ (P$_{MM'}$) = -47.6 ppm, δ (P$_X$) = -163.5 ppm, $^1J_{AA'}$ = 400.5 Hz, $^1J_{AM}$ = $^1J_{A'M'}$ = 409.3 Hz, $^2J_{MM'}$ = -17.7 Hz, $^2J_{AM'}$ = $^2J_{A'M}$ = -37.0 Hz, $^1J_{SIM}$ = $^1J_{SIM'}$ = 169.3 Hz, $^1J_{SIX}$ = 118.3 Hz, $^2J_{SIA}$ = $^2J_{SIA'}$ = 23.7 Hz, [Fe] = Cp*Fe, [Si] = PhC(NtBu)$_2$Si.

69

Table 3.3.4: Chemical shifts, couplings constants and linewidths from the iterative fit of the AA'MM'X spin system of **17** at 298 K.

Parameters	Iteration values	Parameters	Iteration values
$\delta\,(P_A) = \delta\,(P_{A'})$	80.77 ppm	$^1J_{AA'}$	400.5 Hz
$\delta\,(P_M) = \delta\,(P_{M'})$	-47.59 ppm	$^1J_{AM} = {}^1J_{A'M'}$	409.3 Hz
$\delta\,(P_X)$	-163.49 ppm (fragment without Si) / -163.51 ppm (fragment with Si)	$^2J_{MM'}$	-17.7 Hz
$\omega_{1/2}\,(A) = \omega_{1/2}\,(A')$	8.64 Hz	$^2J_{AM'} = {}^2J_{A'M}$	-37.0 Hz
$\omega_{1/2}\,(M) = \omega_{1/2}\,(M')$	8.58 Hz	$^1J_{SiM} = {}^1J_{SiM'}$	169.3 Hz
$\omega_{1/2}\,(X)$	4.90 Hz	$^1J_{SiX}$	118.3 Hz
		$^2J_{SiA} = {}^2J_{SiA'}$	23.7 Hz

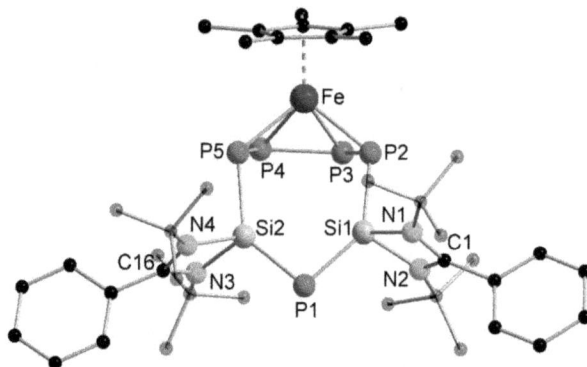

Figure 3.3.15: Molecular structure of **17** in the solid state. H atoms and non-coordinating solvent molecules are omitted for clarity. Selected bond distances (Å) and angles [°]: Fe-P2 2.2799(7), Fe-P3 2.3111(7), Fe-P4 2.3442(6), Fe-P5 2.2689(6), P1-Si1 2.1419(8), P1-Si2 2.1407(9), P2-Si1 2.2151(8), P5-Si2 2.2126(9), P2-P3 2.1416(9), P3-P4 2.1612(9), P4-P5 2.1454(9), Si1-N1 1.8376(2), Si1-N2 1.8415(2), Si2-N3 1.8388(2), Si2-N4 1.850(2), N1-C1 1.332(3), N2-C1 1.340(3), N4-C16 1.330(3); Si2-P1-Si1 94.42(3), P3-P2-Si1 105.57(3), P2-P3-P4 107.59(3), P5-P4-P3 108.20(3), P4-P5-Si2 109.93(3), P1-Si1-P2 134.92(4), N1-C1-N2 106.14(2), N4-C16-N3 105.83(2).

The P-P bond lengths (2.142-2.161 Å) in **17** are also in the same range as those in **14** (2.139-2.160 Å), indicating a partial double bond character. The ring conformation in **17** is similar to the half-chair conformation of cyclohexane. Furthermore, the Si-P seven-membered ring is isoelectronic to the $C_7H_7^-$ tropylium anion.

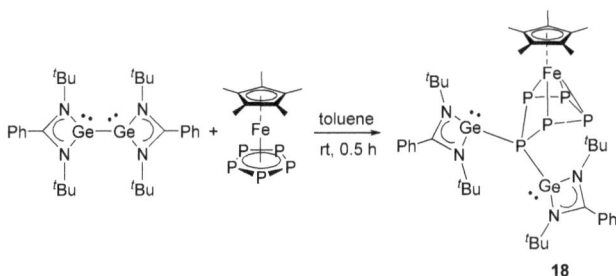

Scheme 3.3.9: Synthesis of complex **18**.

The very unique insertion of [LSi]$^+$ in the P-P bonds of the *cyclo*-P$_5$ ring of [Cp*Fe(η^5-P$_5$)] made us wonder whether this type of ring expansion reactivity could be extended to the heavier group 14 analogues. Nagendran *et al.* have reported the synthesis of [LGe-GeL], the germanium analogue of di-silylene.[181] However, the chemistry of [LGe-GeL] is far less explored in comparison to that of [LSi-SiL], which may be due to the lower reactivity of [LGe-GeL] in terms of reduction potential. In principle, this difference in reactivity could be used to trap reactive intermediates similar to those involved in reactions with [LSi-SiL]. Therefore, we started to investigate the reactivity pattern of [LGe-GeL] towards [Cp*Fe(η^5-P$_5$)]. The reaction between [LGe-GeL] and [Cp*Fe(η^5-P$_5$)] at room temperature in toluene for 12 hours led to the formation of a mixture of products, as evidenced by NMR monitoring of the reaction. Interestingly, the intensity of the different signals in both the ^{1}H and ^{31}P{^{1}H} NMR spectra changed over time. The relatively slow kinetic observed when using [LGe-GeL] potentially allows the isolation of reaction intermediates. In this regard, the reaction between [LGe-GeL] and [Cp*Fe(η^5-P$_5$)] was started at -78 °C and slowly warmed up to room temperature over 30 minutes. After removal of all the volatiles, the residue was quickly extracted with hexane, filtered and stored at -30 °C for one day, resulting in the formation of crystals of [(LGe)$_2${(μ-η^4-P$_5$)FeCp*}] (**18**) (Scheme 3.3.9) in 60% yield. The ^{1}H NMR spectrum of

71

complex **18** shows two singlets at δ 1.23 and 1.32 ppm corresponding to ^{t}Bu groups on the amidinate ligands suggesting their different environments. The signal corresponding to the methyl protons of the Cp* groups is downfield shifted to δ 1.90 ppm, showing the similar trend as observed for complex **14**, **16**, and **17**. The ^{31}P{^{1}H} NMR spectrum showed an AMM'XX' spin system corresponding to three magnetically non-equivalent P atoms with chemical shifts at δ - 45.7 (P$_{XX'}$), 43.5 (P$_{MM}$), and 150.5 (P$_A$), indicating the formation of envelope shape of the cyclo-P$_5$ ring. The different coupling constants were extracted by iterative line fitting of the ^{31}P{^{1}H} NMR spectrum. The values of J_{PP} coupling constants are in usual range and similar to those observed for complexes containing an envelope shaped cyclo-P$_5$ ring such as complex **13** and **16**. The solid-state structure of complex **18** was determined by X-ray crystallography. Complex **18** crystallizes in triclinic space group $P\bar{1}$ with one molecule in an asymmetric unit cell. The molecular structure of complex **18** further confirmed the formation of an envelope shaped cyclo-P$_5$ ring η^4-coordinated to [Cp*Fe]$^+$ fragment (Figure 3.3.17). The Fe-P bond distances (2.298(2)-2.283(2) Å) are in the usual range of the envelope shaped cyclo-P$_5$ ring η^4-coordinated to [Cp*Fe]$^+$ moiety.[81- 82,84] The Ge1-P1 (2.4708(14) Å) and Ge2-P1 (2.419(2) Å) bond lengths are similar to other tri-coordinated germylenes in a polyphosphide environment, for example, [RGe-(μ-P$_2$)-GeR][185] (2.439(1) Å) (**R** = [(p-tolyl)$_2$B{1-(1-adamantyl)-3-yl-2-ylidene}$_2$]) and [P$_7$Ge(N(SiMe$_3$)$_2$)]$^{2-}$ (2.504(1) and 2.526(1) Å).[186] The acute angles between N-Ge-P1 atoms are in the usual range from 94.63(12) to 97.42(12) ° for tricoordinate germylenes.[187] The two germylene moieties, [LGe]$^+$, in complex **18** are in a *trans*-conformation with respect to the orientation of the lone-pairs on Ge centres. Complex **18** is a rare example of germylene in coordination sphere of polyphosphide.[185- 186,188] The very interesting feature of complex **18** is the coordination of both germylene [LGe]$^+$ moieties to one P atom. Isolation of complex **18** gives some insight in the formation of the seven membered ring in complex **17**. The possible intermediate during formation of complex **17** might be analogous to complex **18**. Interestingly, always one equivalent of [LGe-GeL] is reacting with one equivalent of [Cp*Fe(η^5-P$_5$)], even when the reaction is carried out in 1:2 molar ratio, respectively, resulting in complex **18** and one equivalent of unreacted [Cp*Fe(η^5-P$_5$)].

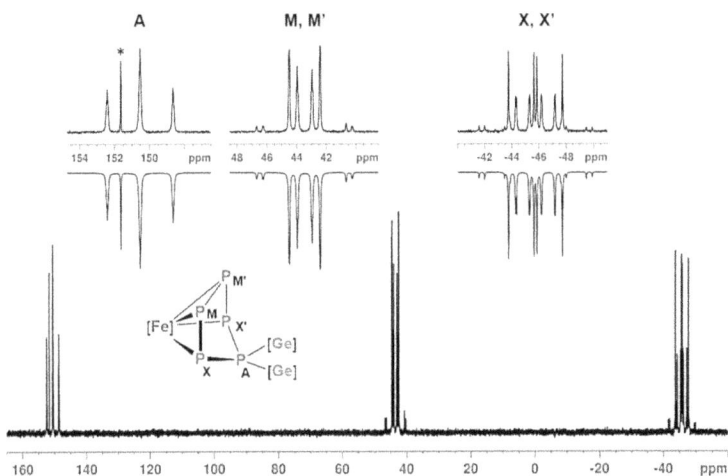

Figure 3.3.16: $^{31}P\{^1H\}$ NMR spectrum at 298 K of compound **18** in C_6D_6 with nuclei assigned to an AMM'XX' spin system; insets: extended signals (upwards) and simulations (downwards); δ (P_A) = 150.5 ppm, δ ($P_{MM'}$) = 43.5 ppm, δ ($P_{XX'}$) = -45.7 ppm, $^1J_{MM'}$ = 406.8 Hz, $^1J_{MX} = {}^1J_{M'X'}$ = 367.2 Hz, $^1J_{AX} = {}^1J_{AX'}$ = 304.6 Hz, $^2J_{XX'}$ = -39.0 Hz, $^2J_{MX'} = {}^2J_{M'X}$ = -27.9 Hz, $^2J_{AM} = {}^2J_{AM'}$ = 8.5 Hz, [Fe] = [Cp*Fe], [Ge] = [LGe] The singlet at δ 151.64 ppm (*) corresponds to trace amounts of [Cp*Fe(η^5-P$_5$)]. Two fragments: complex **18** (statistical weight 1.00) and [Cp*Fe(η^5-P$_5$)] at δ 151.65 ppm (statistical weight 0.02).

Table 3.3.5: Chemical shifts, couplings constants and linewidths from the iterative fit of the AMM'XX' spin system of **18** in C_6D_6 at 298 K.

Parameters	Iteration values	Parameters	Iteration values
δ (P_A)	150.53 ppm	$^1J_{MM'}$	406.8 Hz
δ (P_M) = δ ($P_{M'}$)	43.46 ppm	$^1J_{MX} = {}^1J_{M'X'}$	367.2 Hz
δ (P_X) = δ ($P_{X'}$)	-45.72 ppm	$^1J_{AX} = {}^1J_{AX'}$	304.6 Hz
$\omega_{1/2}$ (Λ)	9.7 Hz	$^2J_{XX'}$	-39.0 Hz
$\omega_{1/2}$ (M) = $\omega_{1/2}$ (M')	8.5 Hz	$^2J_{MX'} = {}^2J_{M'X}$	-27.9 Hz
$\omega_{1/2}$ (X) = $\omega_{1/2}$ (X')	8.8 Hz	$^2J_{AM} = {}^2J_{AM'}$	8.5 Hz

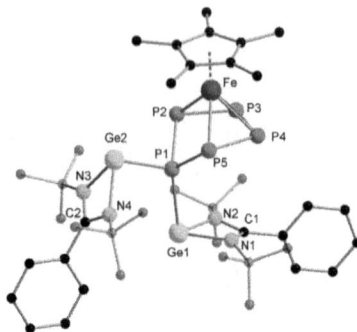

Figure 3.3.17: The molecular structure of complex **18** in the solid state. The hydrogen atoms and the solvent molecule in the unit cell are omitted for clarity. Selected bond distances (Å) and angles [°]: Ge1-P1 2.4708(14), Ge1-N1 1.998(4), Ge1-N2 1.991(4), Ge2-P1 2.419(2), Ge2-N3 2.002(4), Ge2-N4 2.014(4), Fe-P2 2.298(2), Fe-P5 2.283(2), P1-P2 2.182(2), P1-P5 2.178(2), P2-P3 2.160(2), P3-P4 2.134(3), P4-P5 2.133(2), N1-C1 1.334(5), N2-C1 1.343(6), N3-C2 1.333(6), N4-C2 1.342(6); N2-Ge1-N1 65.4(2), N3-Ge2-N4 65.21(2), N1-Ge1-P1 95.78(11), N2-Ge1-P1 97.42(12), N3-Ge2-P1 95.50(11), N4-Ge2-P1 94.63(12), P3-P2-P1 101.33(8), P4-P3-P2 103.34(9), P5-P4-P3 102.80(9).

Complex **18** is not stable in solution at room temperature. However, by careful control of the reaction time, complex **18** could be isolated in a pure form. In order to isolate the other product(s), an NMR-scale reaction was carried out in C_6D_6 at room temperature and the reaction was monitored by ^1H and $^{31}P\{^1H\}$ NMR. As a result, a clean conversion of all the starting materials into complex **18** was observed within 30 minutes. Longer reaction times led to the appearance

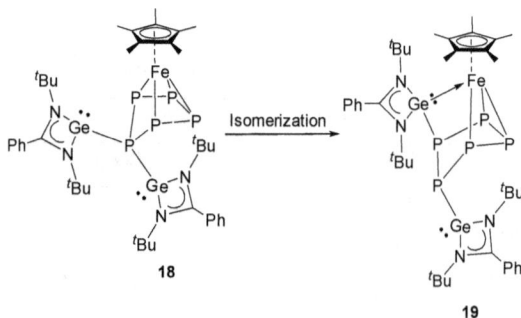

Scheme 3.3.10: Isomerization of complex **18** to complex **19** via 1,2-migration of [LGe]+ moiety.

of new signals in both the ^1H and $^{31}P\{^1H\}$ NMR spectra, indicating transformation of complex **18** into new species along with reformation of a small amount of [Cp*Fe(η^5-P$_5$)]. Although an increase of the signals corresponding to the degradation product of complex **18** was observed at room temperature, the rate of the reaction was very slow which prevented full conversion in the NMR tube, even after one month at room temperature. To increase the rate of the reaction, [LGe-GeL] and [Cp*Fe(η^5-P$_5$)] were heated in toluene at 60 °C for 3 hours. As a result, after a short work-up, a mixture of crystals was obtained, consisting mostly of [Cp*Fe(η^5-P$_5$)] along with a few yellow-coloured crystals of [(LGe){(μ-η^3-P$_5$)(η^1-GeL)FeCp*}] (**19**) as identified by X-ray diffraction studies (Scheme 3.3.10 and Figure 3.3.18). Complex **19** is an isomer of complex **18** that is formed by an unprecedented 1,2-migration of one [LGe]$^+$ moiety on the *cyclo*-P$_5$ ring. The migrated [LGe]$^+$ moiety further inserts into one of the Fe-P bonds, resulting in an unusual η^3-coordination of the *cyclo*-P$_5$ ring to the [Cp*Fe]$^+$ moiety. The Fe-P5 (2.3311(11) Å) and Fe-P4 (2.3402(10) Å) bond distances in complex **19** are similar to the Fe-P bond lengths in **18**. On the other hand, the Fe-P3 (2.3936(10) Å) bond distance is significantly longer, which could be attributed to the insertion of the [LGe]$^+$ moiety into the adjacent Fe-P2 bond. The Ge1-P1 bond length in complex **19** is shorter than the corresponding bond distance in complex **18** (2.4299(9) *vs* 2.4708(14) Å, respectively). The distance of the newly formed Ge2-P2 bond (2.3131(10) Å) is on the shorter end of the reported Ge-P bond lengths. Such a short Ge-P separation might be the result of the ring strain imposed by the four-membered Ge2-Fe-P3-P2 ring and also the decrease in the electron density at the Ge2 atom upon donation of its lone pairs to the Fe atom. The P1-P2 (2.1899(12) Å), P3-P4 (2.1197(12) Å), and P4-P5 (2.1600(11) Å) bond distances are intermediate between P-P single and double bonds.[174,176,178a,178b] The P1-P5 (2.2123(12) Å) and P2-P3 (2.3189(12) Å) distances are consistent with a single bond and a weak single bond, respectively.[174,176,178a,178b] The elongation of the P1-P5 and P2-P3 bonds at the 1,3 positions of the *cyclo*-P$_5$ ring in complex **19** is very interesting, as the insertion of the [LSi]$^+$ moieties in complex **17** occurred at same positions. Therefore, complex **19** might be structurally similar to the intermediate leading to the ring expansion of the *cyclo*-P$_5$ ring in the reaction of [Cp*Fe(η^5-P$_5$)] with [LSi-SiL]. Despite several attempts, analytically pure complex **19** could not be isolated due to two main reasons: i) if the reaction is carried out at room temperature, full conversion of complex **18** to complex **19** is impossible in a reasonable time scale (more than one month); ii) upon heating the reaction mixture to increase the rate of the reaction (60 °C for 3 hours), thermal decomposition is observed, resulting in a mixture of starting material ([Cp*Fe(η^5-P$_5$)]) along with a small amount of crystals of complex **19** and an unidentified precipitate. Nevertheless, NMR analysis of a mixture of **18**, **19** and [Cp*Fe(η^5-P$_5$)], obtained after reaction of [LGe-GeL] with [Cp*Fe(η^5-P$_5$)] in C$_6$D$_6$ for one month, allowed the successful extraction of the ^{1}H and ^{31}P{^{1}H} NMR signals of complex **19** by comparison with the signals of isolated **18** and ([Cp*Fe(η^5-P$_5$)] (Figure 3.3.19 and 3.3.20). The ^{1}H NMR spectrum showed a very slight downfield shift of the Cp* methyl protons upon isomerization of **18** to **19** (δ 1.90 vs 1.91, respectively) (Figure 3.3.19). In addition, four new signals at δ 1.10, 1.31, 1.42, and 1.52 were detected in the aliphatic region, corresponding to the ^{t}Bu groups of complex **19**. Notably, the presence of four resonance indicates the non-equivalence of the ^{t}Bu groups, which is possibly due to a rigid conformation and restricted rotation of the amidinate ligand. In agreement with the solid-state structure of **19**, the ^{31}P{^{1}H} NMR spectrum showed a set of five multiplets at δ -43.01 (P$_5$), -6.84 (P$_4$), 72.42 (P$_3$), 90.43 (P$_2$), and 252.54 (P$_1$) ppm, corresponding to five non-equivalent P atoms. The different coupling constants were successfully extracted by iterative simulations of the spectrum (Figure 3.3.21 and Table 3.3.6).

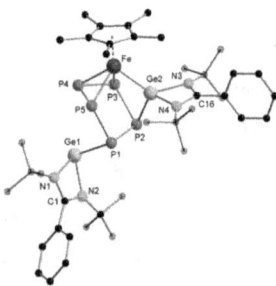

Figure 3.3.18: The molecular structure of complex **19** in the solid state. The hydrogen atoms and the solvent molecule in the unit cell are omitted for clarity. Selected bond distances (Å) and angles [°]: Fe-Ge2 2.2747(8), Ge1-P1 2.4299(9), Ge1-N1 2.021(2), Ge1-N2 2.029(2), Ge2-P2 2.3131(10), Ge2-N3 2.004(2), Ge2-N4 1.994(2), Fe-P3 2.3936(10), Fe-P4 2.3402(10), Fe-P5 2.3311(11), P1-P2 2.1899(12), P1-P5 2.2123(13), P2-P3 2.3189(12), P3-P4 2.1197(12), P4-P5 2.1600(11), N1-C1 1.323(3), N2-C1 1.328(3), N3-C16 1.332(3), N4-C16 1.327(3); N1-Ge1-P1 96.42(6), N2-Ge1-P1 95.22(7), N3-Ge2-P2 108.20(7), N4 Ge2 P2 117.21(7), N1-C1-N2 109.3(2), N4-C16-N3 109.8(2).

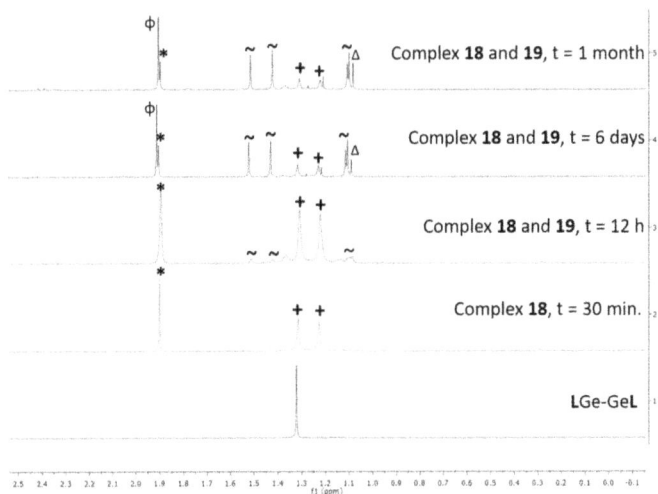

Figure 3.3.19: Stacked ^{1}H NMR spectra of the aliphatic region of the formation of complex **18** and its conversion to complex **19**.

Part 1 of the picture shows the ^{t}Bu group of [LGe-GeL]. + = ^{t}Bu group of **18**, * = Cp* group of **18**, ~ = ^{t}Bu group of **19**, φ = Cp* group of **19**, Δ = [Cp*Fe(η^5-P$_5$)].

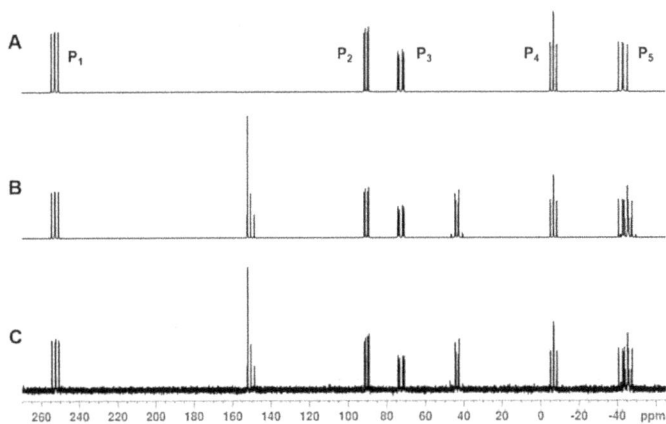

Figure 3.3.20: ^{31}P{^{1}H} NMR spectrum of compound **19** (containing complex **18** and [Cp*Fe(η^5-P$_5$)]) recorded in C$_6$D$_6$ at 298 (**C**), and simulation of all the species in solution (*i.e.* **19**, **18**, and [Cp*Fe(η^5-P$_5$)] with the statistical weight 1.00, 0.42 and 0.37, respectively) (**B**), and simulation spectrum including only the **19** product (**A**).

77

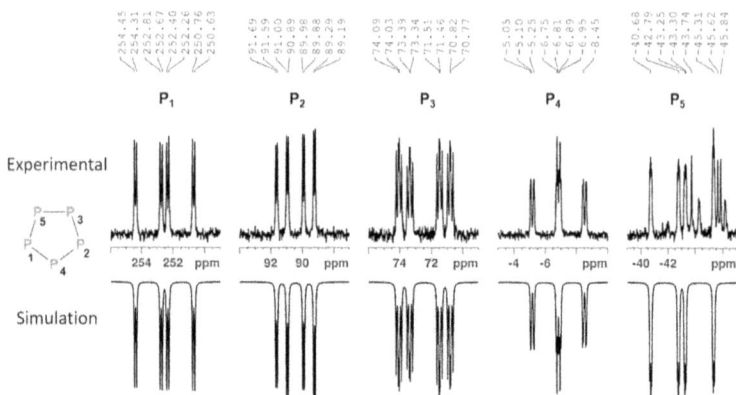

Figure 3.3.21: Section of the ^{31}P{^{1}H} NMR spectrum (162.06 MHz, 298 K, C_6D_6) of compound **19** with assignment of the nuclei; experimental (upwards) and simulation (downwards).

Table 3.3.6: Chemical shifts, couplings constants and linewidths from the iterative fit of the P_5 spin system of **19** in C_6D_6 at 298 K. Probably Ge coordinated to P3 and P5 because larger value for the linewidth ($\omega_{1/2}$).

Parameters	Iteration values	Parameters	Iteration values
δ (P_1)	252.54 ppm	$^1J_{P1-P4}$	265.7 Hz
δ (P_2)	90.43 ppm	$^1J_{P1-P5}$	333.3 Hz
δ (P_3)	72.42 ppm	$^1J_{P2-P3}$	112.1 Hz
δ (P_4)	-6.84 ppm	$^1J_{P2-P4}$	277.1 Hz
δ (P_5)	-43.01 ppm	$^1J_{P3-P5}$	416.2 Hz
$\omega_{1/2}$ (P_1)	6.4 Hz	$^2J_{P1-P2}$	-4.2 Hz
$\omega_{1/2}$ (P_2)	6.4 Hz	$^2J_{P1-P3}$	22.1 Hz
$\omega_{1/2}$ (P_3)	7.4 Hz	$^2J_{P2-P5}$	17.2 Hz
$\omega_{1/2}$ (P_4)	6.1 Hz	$^2J_{P3-P4}$	31.9 Hz
$\omega_{1/2}$ (P_5)	7.3 Hz	$^2J_{P4-P5}$	8.8 Hz

3.3.6 Reactivity of [Cp*Fe(η^5-P$_5$)] with mono-valent magnesium complex

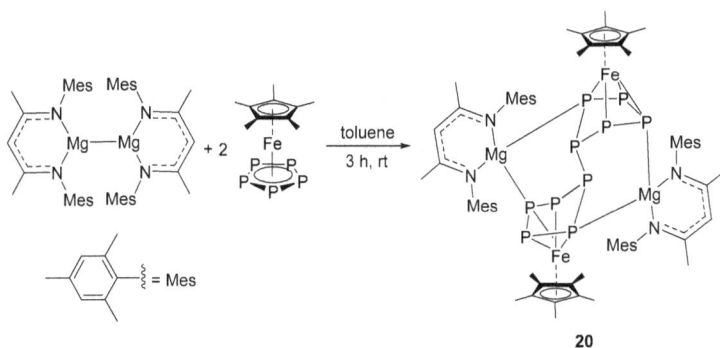

20

Scheme 3.3.11: Synthesis of complex 20.

In 2007, Jones and co-workers reported a series of molecular mono-valent magnesium complexes.[189] Apart from being a fundamental breakthrough, Mg^I complexes featuring a Mg-Mg bond have shown potential applications due to their strong reducing nature.[190] The selectivity of Mg^I compounds in reduction reactions has rendered them a promising class of reductants,[191] and they have been promoted as "quasi-universal" reducing agents.[34] Mg^I complexes act as single-electron reductants and the Mg centre is oxidized from the +1 to the more stable +2 oxidation state. In many cases, the reactivity of Mg^I complexes is similar to that of low-valent f-block elements. For example, the reduction of CO by an activated Mg^I complex resulted in the formation of the [C$_3$O$_3$]$^{3-}$ deltate anion, which can also be obtained by reduction of CO with a trivalent uranium complex.[192] As described in section 1.4.2, Roesky, Scheer and co-workers have been investigating the behaviour of [Cp*Fe(η^5-P$_5$)] upon reaction with single-electron reductants such as elemental potassium and divalent samarium complexes. In the previous sections, we have shown that, by using different reducing agents, different types of polyphosphides could be isolated. Comparison of the reactivity of [Mes-*BDI*-Mg^I]$_2$[34,189] (Mes-*BDI* = (2,4,6-Me$_3$C$_6$H$_3$NCMe)$_2$CH) with that of [LGe-GeL] in the reduction of [Cp*Fe(η^5-P$_5$)] is especially relevant as, in case of [LGe-GeL], the [P$_{10}$]$^{4-}$ fragment formed by the formal single-electron reduction of [Cp*Fe(η^5-P$_5$)] could not be observed.

The reaction of [Mes-*BDI*-Mg^I]$_2$[193] with [Cp*Fe(η^5-P$_5$)] in a 1:2 molar ratio, respectively, resulted in the formation of [(Mes-*BDI*-Mg^{II})$_2$(μ-η^4-η^4-P$_{10}$)(FeCp*)$_2$] (**20**) (Scheme 3.3.11). The solid-state structure of complex **20** was determined by X-ray diffraction studies. Complex **20** crystallizes in the orthorhombic space group *Pbcn* with one molecule in the asymmetric unit cell. The molecular structure showed the formation of a [P$_{10}$]$^{4-}$ fragment, η^4-coordinated to two [Cp*Fe]$^+$ and η^2-coordinated to two [Mes-*BDI*-Mg^{II}]$^+$ moieties (Figure 3.3.22). Similar [(η^4-η^4-P$_{10}$)(FeCp*)$_2$]$^{2-}$ moieties have been obtained coordinated to potassium and samarium ions upon reduction of [Cp*Fe(η^5-P$_5$)] by potassium hydride[81] and divalent samarocene,[83] respectively. The Mg1-P9 (2.614(2) Å), Mg1-P2 (2.578(2) Å), Mg2-P4 (2.643(2) Å), and Mg2-P7 (2.590(2) Å) bond lengths are similar to those observed in previously reported magnesium polyphosphide complexes (2.5592(7)-2.831(3) Å).[194] The P10-Mg1 (3.192(2) Å) and P3-Mg2 (3.140(2) Å) separations are however significantly longer than the Mg-P bonds reported in the literature (2.45-2.99 Å), precluding a bonding interaction between P10 and Mg1 on the one hand, and P3 and Mg2 on the other hand.[195] The Fe-P bond distances are ranging from 2.276(2) to 2.332(2) Å, and are slightly shorter than those observed in the related samarium complex [(Cp*$_2$Sm)$_2$(μ-η^4-η^4-P$_{10}$)(FeCp*)$_2$] (2.303(11)-2.352(12) Å).[83] The P-P bond lengths (2.154(2)-2.174(2) Å) within the *cyclo*-P$_5$ ring are intermediate between single and double bonds, indicating a partial double-bond character.[174] The distance of the newly formed P1-P6 bond (2.198(2) Å) in complex **20** is also slightly shorter than the analogous P-P bond in [(Cp*$_2$Sm)$_2$(μ-η^4-η^4-P$_{10}$)(FeCp*)$_2$] (2.2089(13) Å). The formation of complex **20** can be explained by the transfer of two electrons from the Mg^I dimer to two molecules of [Cp*Fe(η^5-P$_5$)]. As a result, two 19e$^-$ iron species with a radical located on one P atom of the *cyclo*-P$_5$ ring are generated *in situ* and subsequently dimerize through the coupling of the radicals and the formation of a P-P bond. This transformation is accompanied by a conformational change of the *cyclo*-P$_5$ rings, from planar to envelope-shaped.[81,83] To best of our knowledge, complex **20** represents the largest structurally characterized polyphosphide in the coordination sphere of magnesium ions. The ^{1}H NMR spectrum of complex **20** shows a downfield shift for the protons of the Cp* group, from δ 1.08 to 1.21 ppm, compared to [Cp*Fe(η^5-P$_5$)]. The protons of the Mes-*BDI* ligands coordinated to the magnesium centres are also downfield shifted. For example, the o-CH$_3$ of the mesityl group are shifted from δ 1.91 ppm in [Mes-*BDI*-Mg^I]$_2$ to

δ 2.53 ppm in complex **20**.[193] The $^{31}P\{^1H\}$ NMR spectrum confirms the formation of a $[P_{10}]^{4-}$ moiety by the presence of four resonances at δ -30.6 (br, $\Delta\nu_{1/2} \approx$ 600 Hz), 24.5 (br, $\Delta\nu_{1/2} \approx$ 550 Hz), 126.5 (br, $\Delta\nu_{1/2} \approx$ 500 Hz), and 152.1 (br, $\Delta\nu_{1/2} \approx$ 500 Hz) ppm. Unfortunately, the P-P coupling constants could not be determined as finely-resolved $^{31}P\{^1H\}$ NMR spectra could not be obtained even at low temperature (-80 °C) due to fluxional behaviour of the $[P_{10}]^{4-}$ fragment. Complex **20** slowly decomposes to unidentified products at room temperature, both in solution and in the solid-state. However, no decomposition was observed when stored at -40 °C for at least a few weeks.

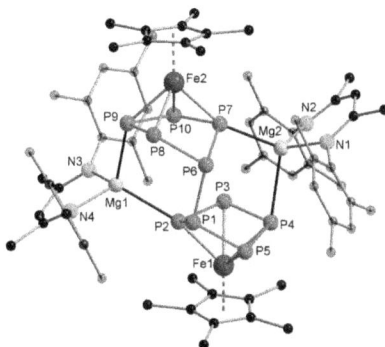

Figure 3.3.22: The molecular structure of complex **20** in solid state. Hydrogen atoms are omitted for clarity. Selected bond distances (Å) and angles [°]: Fe1-P2 2.276(2), Fe1-P3 2.329(2), Fe1-P4 2.301(2), Fe1-P5 2.332(2), Fe2-P7 2.284(2), Fe2-P8 2.327(2), Fe2-P9 2.293(2), Fe2-P10 2.328(2), P1-P2 2.165(2), P2-P3 2.156(2), P3-P4 2.175(2), P4-P5 2.162(3), P5-P1 2.173(2), P1-P6 2.197(2), P6-P7 2.166(2), P6-P8 2.173(2), P7-P10 2.152(2), P8-P9 2.160(3), P9-P10 2.167(2), P2-Mg1 2.580(2), P9-Mg1 2.614(2), P10-Mg1 3.192(2), P4-Mg2 2.643(2), P7-Mg2 2.590(2), P3-Mg2 3.140(2); P2-P1-P5 90.22(8), P2-P1-P6 112.62(8), P5-P1-P6 108.52(9), P3-P2-P1 111.96(9), P2-P3-P4 99.57(9), P5-P4-P3 104.48(9), P4-P5-P1 108.42(9), P7-P6-P1 113.09(8), P7-P6-P8 90.11(8), P8-P6-P1 108.11(9), P10-P7-P6 111.69(9), P9-P8-P6 108.44(9), P8-P9-P10 104.39(9), P7-P10-P9 99.84(9).

3.3.7 Reactivity of [Cp*Fe(η^5-P$_5$)] with monovalent aluminium complexes

Monovalent aluminium complexes are considered as group-13 analogues of carbenes but their reactivity pattern is different in some cases.[196] After the functionalization of [Cp*Fe(η^5-P$_5$)] with group-14 elements, we planned to study the reactivity of [Cp*Fe(η^5-P$_5$)] with monovalent aluminium complexes. The reaction between equimolar amounts of [Cp*Fe(η^5-P$_5$)] and [Dipp-BDI-Al^I] in toluene at room temperature resulted in the formation of [(Dipp-BDI-Al^{III})(μ-η^3-η^4-P$_5$)FeCp*] (**21**), isolated in 35% yield (Scheme 3.3.12). NMR spectrum of **21** showed that the resonance for the methyl protons of the Cp* moiety is shifted downfield from δ 1.08 ppm in [Cp*Fe(η^5-P$_5$)] to δ 1.22 ppm. Considerable shifts of the signals are also observed for the protons of the Dipp-BDI group coordinated to aluminium, in particular the resonances of the ^{i}Pr groups which are shifted from δ 1.13 and 1.38 ppm in [Dipp-BDI-Al^I] to δ 1.04 and 1.93 ppm in **21**. Also, three new broad resonances at δ 98.4, 60.7, and 32.1 ppm were observed in the ^{31}P{^{1}H} NMR spectrum at room temperature, suggesting a fluxional behaviour of the *cyclo*-P$_5$ ring. A well-resolved ^{31}P{^{1}H} NMR spectrum could be recorded at 233 K, showing an AA'MXX' spin system with multiplets at δ 97.2 (P$_{AA'}$), 60.6 (P$_M$), and 30.7 (P$_{XX'}$) ppm consistent with the formation of an envelope conformation for the *cyclo*-P$_5$ ring. The multiplets were assigned by higher order P-P coupling patterns and the coupling constants were obtained from an iterative fitting of the ^{31}P{^{1}H} NMR spectrum (Figures 3.3.23 and Table 3.3.7). The conformational change of the *cyclo*-P$_5$ ring was further confirmed by the molecular structure of **21**, retrieved by X-ray crystallographic analysis and confirming the formation of a triple-decker type complex with a bent *cyclo*-P$_5$ ring (Figure 3.3.24). The envelope-shaped *cyclo*-P$_5$ ring is η^4-coordinated to the [Cp*Fe]$^+$ moiety, η^3-coordinated to the [Dipp-BDI-Al^{III}]$^{2+}$ fragment, and acts as a bridge between the two separate halves of the triple-decker complex. To the best of our knowledge, **21** is the first example of an Al containing neutral heterometallic triple-decker complex. The average Fe-P bond length in complex **21** (2.273 Å) is slightly shorter than the average Fe-P bond length in [{K(dme)K(dibenzo[18]crown-6)}(Cp*Fe(η^5-P$_5$))] (2.291 Å).[81] The Al-P1 (2.323(2) Å) and Al-P4 (2.465(2) Å) bond distances are in the reported range of Al-P single bonds (2.308(2) to 2.422(2) Å). In contrast, the relatively longer Al-P5 (2.784(2) Å) bond distance indicates only a weak coordination between Al and P5.[119] The P2-P3 (2.1647(14) Å) and P3-P4 (2.186(2) Å) bond lengths are shorter than the P1-P2 (2.223(2) Å), P1-P5 (2.2145(14) Å) and P4-P5 (2.2735(14) Å) analogues, which can be ascribed to the coordination of the cyclo-P_5 ring to the Lewis acidic aluminium atom in complex **21**. A similar trend has been observed in a related samarium polyphosphide complex.[84] The reduction of [Cp*Fe(η^5-P$_5$)] by [Dipp-BDI-Al^I] complex is in sharp contrast with the reported reactivity of [Cp*Fe(η^5-P$_5$)] with cationic [Ga^I]$^+$ and [Tl^I]$^+$ species. In the case of [Tl^I(Al{OC(CF$_3$)$_3$}$_4$)] and [Ga^I(Al{OC(CF$_3$)$_3$}$_4$)], only coordination polymers featuring a planar cyclo-P_5 ring, [{Tl^I(Al{OC(CF$_3$)$_3$}$_4$)}(μ-η^5-η^5-η^1-P$_5$)FeCp*] and [{Ga^I(Al{OC(CF$_3$)$_3$}$_4$)}(μ-η^5-η^5-η^1-P$_5$)FeCp*], were obtained.[171a,171g] This anomalous trend in reactivity can be mainly attributed to the higher reductive ability of [Al^I] complexes as compared to the [Ga^I] and [Tl^I] analogues. Interestingly, NMR studies of the reaction between [Dipp-BDI-Al^I] and [Cp*Fe(η^5-P$_5$)] consistently showed the formation of other products and revealed that complex **21** further reacts with another equivalent of [Dipp-BDI-Al^I]. To gain more insight in the nature of this reduction compound, the reaction between [Dipp-BDI-Al^I] and [Cp*Fe(η^5-P$_5$)] was carried out in a 2:1 molar ratio, respectively, and resulted in the clean formation of a new species which was identified by ^{31}P{^{1}H} NMR spectroscopy (Figure 3.3.25). Unfortunately, despite several attempts, no crystals suitable for X-ray diffraction studies could be obtained and the exact identity of this product remains elusive to date.

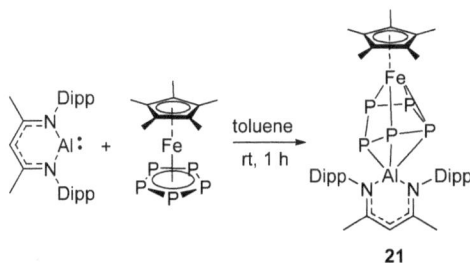

Scheme 3.3.12: Synthesis of complex **21**.

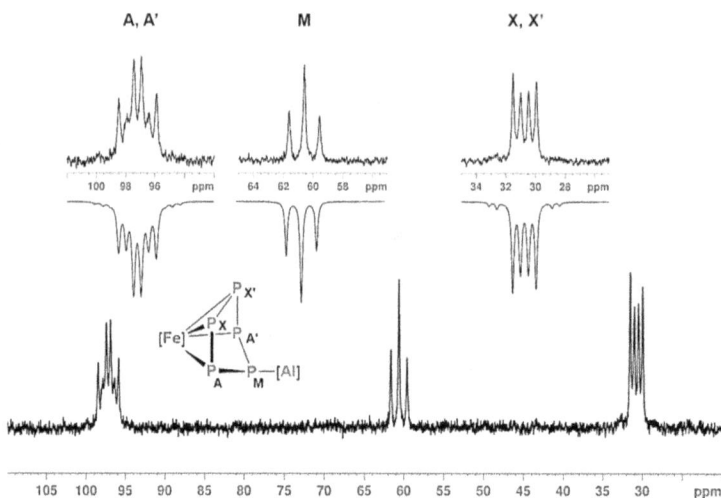

Figure 3.3.23: $^{31}P\{^1H\}$ NMR spectrum (161.97 MHz, 233 K, toluene-d_8) of compound **21** with nuclei assigned to an AA'MXX' spin system; insets: extended signals (upwards) and simulations (downwards); δ (P$_{AA'}$) = 97.2 ppm, δ (P$_M$) = 60.6 ppm, δ (P$_{XX'}$) = 30.7 ppm, $^1J_{XX'}$ = 301.0 Hz, $^1J_{AX}$ = $^1J_{A'X'}$ = 248.2 Hz, $^1J_{MA}$ = $^1J_{MA'}$ = -165.9 Hz, $^2J_{AA'}$ = 48.5 Hz, $^2J_{XA}$ = $^2J_{X'A'}$ = -5.3 Hz, $^2J_{XM}$ = $^2J_{X'M}$ = 3.8 Hz, [Fe] = [Cp*Fe]$^+$, [Al] = [Dipp-*BDI*-Al^{III}]$^{2+}$.

Table 3.3.7: Chemical shifts, couplings constants and linewidths from the iterative fit of the P$_5$ spin system of **21** in toluene-d_8 at 233 K.

Parameters	Iteration values	Parameters	Iteration values
δ (P$_A$) = δ (P$_{A'}$)	97.17 ppm	$^1J_{XX'}$	301.0 Hz
δ (P$_M$)	60.62 ppm	$^1J_{XA}$ = $^1J_{X'A'}$	248.2 Hz
δ (P$_X$) = δ (P$_{X'}$)	30.74 ppm	$^1J_{MA}$ = $^1J_{MA'}$	-165.9 Hz
$\omega_{1/2}$ (A) = $\omega_{1/2}$ (A')	43.2 Hz	$^2J_{AA'}$	48.5 Hz
$\omega_{1/2}$ (M)	29.6 Hz	$^2J_{XA'}$ = $^2J_{X'A}$	-5.3 Hz
$\omega_{1/2}$ (X) = $\omega_{1/2}$ (X')	31.3 Hz	$^2J X_M$ = $^2J_{X'M}$	3.8 Hz

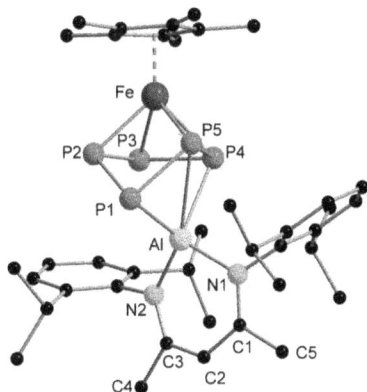

Figure 3.3.24: Molecular structure of **21** in the solid state. H atoms are omitted for clarity. Selected bond distances (Å) and angles [°]: Fe-P2 2.2578(11), Fe-P3 2.2803(12), Fe-P4 2.3403(12), Fe-P5 2.2110(12), Fe-C30 2.117(4), Fe-C31 2.074(4), Fe-C32 2.088(4), Fe-C33 2.113(4), Fe-C34 2.128(4), P1-P2 2.223(2), P1-P5 2.2145(14), P1-Al 2.323(2), P2-P3 2.1647(14), P3-P4 2.186(2), P4-P5 2.2735(14), P4-Al 2.465(2), P5-Al 2.784(2), Al-N1 1.914(3), Al-N2 1.907(3), N1-C1 1.335(5), N2-C3 1.336(5), C1-C2 1.391(5), C1-C5 1.505(5), C2-C3 1.399(5), C3-C4 1.502(5); P2-P1-Al 90.07(5), P5-P1-P2 82.85(5), P5-P1-Al 75.66(5), P3-P2-P1 107.95(6), P2-P3-Fe 60.99(4), P2-P3-P4 100.67(6), P4-P3-Fe 63.16(4), P3-P4-P5 98.59(5), P3-P4-Al 78.85(5), P5-P4-Al,71.84(5), N2-Al-N1 96.12(14).

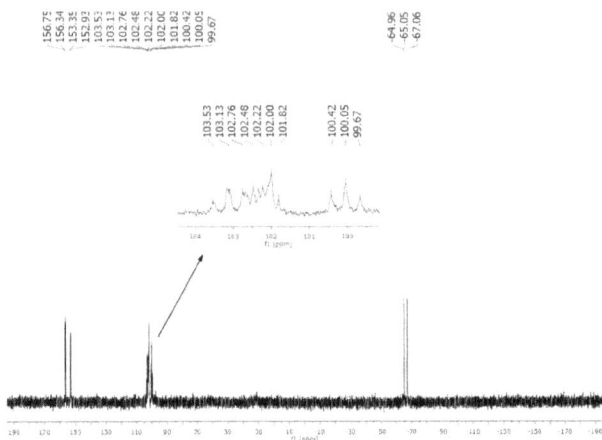

Figure 3.3.25: $^{31}P\{^1H\}$ NMR (162 MHz, 298 K, C_6D_6) spectrum of the reaction between [Dipp-*BDI*-Al^I] and [Cp*Fe(η^5-P$_5$)] in a 2:1 molar ratio, respectively.

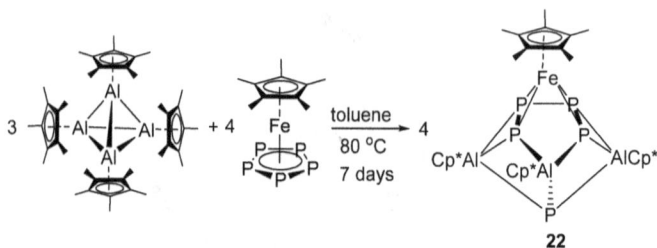

Scheme 3.3.13: Synthesis of complex **22**.

In order to obtain a deeper understanding of the reactivity pattern of $[Cp*Fe(\eta^5-P_5)]$ with monovalent aluminium complexes, another monovalent aluminium complex, namely $[(Cp*Al^I)_4]$, was selected. The reaction between $[(Cp*Al^I)_4]^{[107,110]}$ and $[Cp*Fe(\eta^5-P_5)]$ was carried out in toluene at room temperature in a 1:4 molar ratio, respectively. In the $^{31}P\{^1H\}$ NMR spectrum of the reaction mixture, two new resonances arose at δ 73.4 ppm and δ -202.9 ppm, indicating the formation of a new product. Although the reaction was very slow at room temperature and did not reach completion even after 3 weeks, the rate of reaction could be increased by heating the reaction mixture up to 80 °C until full consumption of $[(Cp*Al^I)_4]$. A small amount of red-coloured crystals suitable for X-ray diffraction studies were grown by slow evaporation of the toluene solvent. The solid-state structure of the corresponding complex showed the formation of $[(P)(Cp*Al^{III})_3(\mu-\eta^2-\eta^2-\eta^2-\eta^4-P_4)(FeCp*)]$ (**22**) featuring an unprecedented tetrametallic Al-Fe polyphosphide core with one iron, five phosphorous and three aluminium atoms (Figure 3.3.26). The crystalline yield of complex **22** could be increased up to 47% by adjusting the stoichiometry between $[(Cp*Al^I)_4]$ and $[Cp*Fe(\eta^5-P_5)]$ to a 3:4 molar ratio, respectively (Scheme 3.3.13). Complex **22** formally results from the 6e$^-$ reduction of one molecule of $[Cp*Fe(\eta^5-P_5)]$ by three $[Cp*Al^I]$ moieties via oxidation of Al from +1 to +3. The insertion of three $[Cp*Al^{III}]^{2+}$ moieties into one P-P bond of the cyclo-P_5 ring results in the formation of a $[cyclic-P_4(Cp*Al)]$ moiety and one P unit. The cleavage of the cyclo-P_5 ring of $[Cp*Fe(\eta^5-P_5)]$ into P_4 and P fragments was also observed in the case of complex **17**, where a formal 6e$^-$ reduction of $[Cp*Fe(\eta^5-P_5)]$ occurred upon insertion of two $[LSi]^+$ moieties into the cyclo-P_5 ring. As shown in figure 3.3.26, the $[cyclic-P_4(Cp*Al)]$ moiety is η^4-coordinated to the $[Cp*Fe]^+$ unit, resulting in an average Fe-P bond length slightly longer than that in [Cp*Fe(η^5-P$_5$)] (2.317 vs 2.273 Å, respectively).[79] The five-membered

[cyclic-P$_4$(Cp*Al)] ring is further coordinated to two [Cp*Al^{III}]$^{2+}$ in an η^2-coordination mode and

to the terminal P fragment. The Al1 and Al2 atoms are η^5-coordinated while Al3 is only η^3-

coordinated by the respective Cp* groups. However, the ^{1}H NMR spectrum of **22** only revealed

two sharp singlets at δ 2.16 ppm (45 H, Cp* on Al atoms) and δ 1.30 ppm (15 H, Cp* on Fe atom),

suggesting the fluxional behaviour in solution of the Cp* moieties bonded to Al. The average Al-

P5 bond length (2.316(2) Å) is in the usual range of Al-P single bonds (from 2.308(2) to 2.422(2)

Å).[119] However, the Al1-P1 (2.476(2) Å), Al1-P2 (2.676(2) Å), Al2-P3 (2.678(2) Å), and Al2-P4

(2.517(2) Å) bond distances are longer than usual Al-P single bonds, which suggests a weaker

coordination. Besides, the short average P-P bond length (2.191(2) Å) in the [cyclic-P$_4$(Cp*Al)]

indicates a partial double-bond character.[174] At room temperature, the ^{31}P{^{1}H} NMR spectrum

of **22** showed only two singlets at δ 73.4 ppm (4 P, acyclic-P$_4$) and δ -202.9 ppm (1 P, terminal P).

Surprisingly, even at low temperature no P-P coupling could be observed, although broadening

of the signals occurred upon decreasing the temperature (Figure 3.3.27).

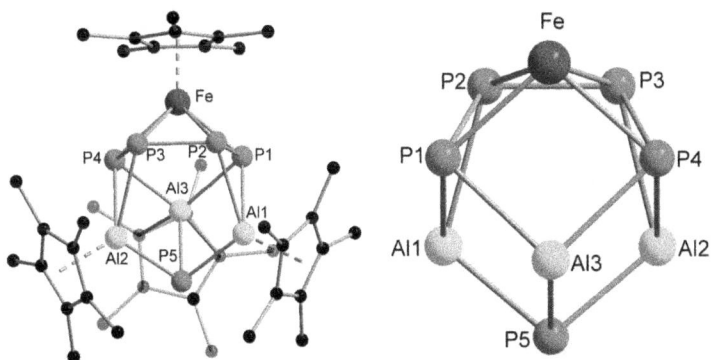

Figure 3.3.26: Molecular structure of **22** (left) in the solid state and simplified view of the core structure of **22** (right) without the Cp* moieties. H atoms are omitted for clarity. Selected bond distances (Å) and angles [°]: Fe-P1 2.336(2), Fe-P2 2.282(2), Fe-P3 2.289(2), Fe-P4 2.361(2), Fe-C31 2.119(5), Fe-C32 2.116(5), Fe-C33 2.087(5), Fe-C34 2.082(5), Fe-C35 2.088(6), P1-P2 2.193(2), P1-Al1 2.476(2), P1-Al3 2.439(2), P2-P3 2.192(2), P2-Al1 2.676(2), P3-P4 2.189(2), P3-Al2 2.678(2), P4-Al2 2.517(2), P4-Al3 2.417(2), P5-Al1 2.309(2), P5-Al2 2.305(2), P5-Al3 2.334(2); P1-Fe-P4 97.82(5), P2-Fe-P1 56.68(5), P2-Fe-P3 57.29(5), P2-Fe-P4 99.50(6), P3-Fe-P1 99.97(5), P3-Fe-P4 56.12(5), P3-P2-P1 107.80(8), P4-P3-P2 108.05(7), P1-Al1-P2 50.21(5), P5-Al1-P1 104.59(7), P5-Al1-P2 112.91(7), P4-Al2-P3 49.71(5), P5-Al2-P3 113.34(8), P5-Al2-P4 103.58(7), P5-Al3-P1 105.01(7), P5-Al3-P1 105.01(7).

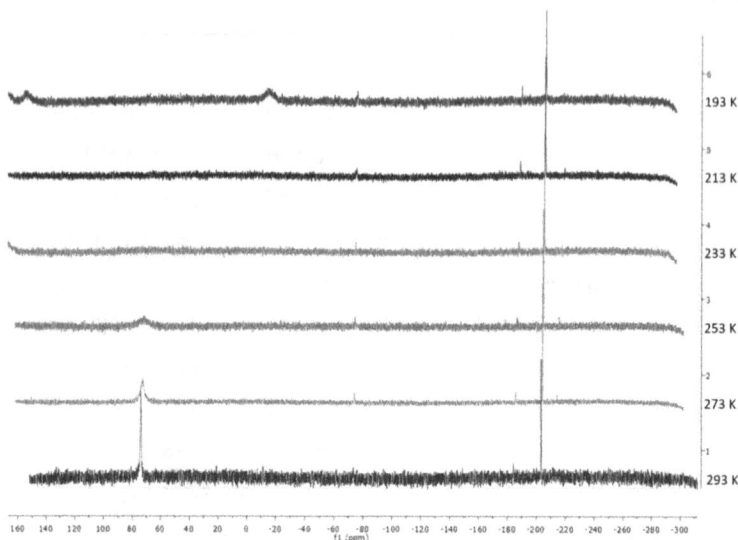

Figure 3.3.27: Variable temperature ^{31}P{^{1}H} NMR (162 MHz, toluene-d_8) spectra of complex **22**.

It should be noted that the exclusive formation of complex **22** was observed when using [(Cp*Al^I)$_4$] as a reducing agent. Taking into consideration that, in the case of [Dipp-*BDI*Al^I], the formally di-reduced product of [Cp*Fe(η^5-P$_5$)] could be isolated, we anticipated that the presence of an additional donor ligand in the reaction of [(Cp*Al^I)$_4$] with [Cp*Fe(η^5-P$_5$)] might stabilize possible reaction intermediates. For that purpose, dimethoxyethane was selected as a suitable ligand candidate, because of its redox-inactivity and its established behaviour as a strong chelating donor ligand for Lewis acidic metals. An NMR-scale reaction between [(Cp*Al^I)$_4$] and [Cp*Fe(η^5-P$_5$)] in a molar ratio of 3:4 was carried out in the presence of dimethoxyethane. Initially, a set of three resonances at δ -122.8, 33.9, and 121.3 ppm was observed in the ^{31}P{^{1}H} NMR spectrum, which may correspond to the possible reaction intermediate **22i** (Figure 3.3.28). These signals disappear after a few hours at room temperature, giving rise to the characteristic signals of complex **22**. The ^{31}P{^{1}H} NMR spectrum of **22i** is consistent with an AA'MM'X spin system, which is a characteristic pattern observed for di-reduced and bent *cyclo*-P$_5$ rings.

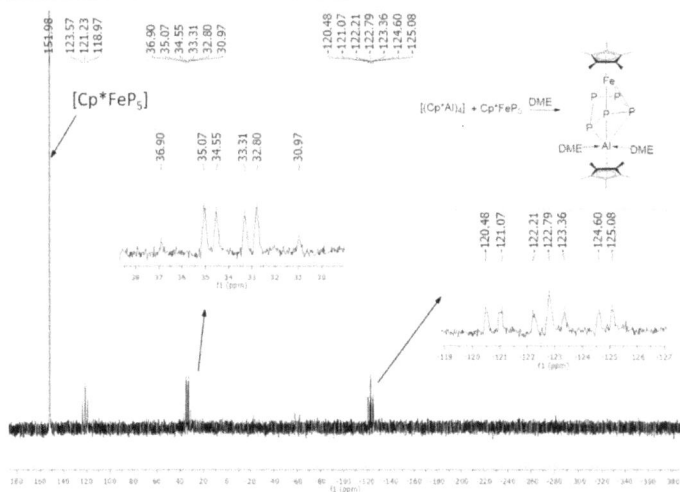

Figure 3.3.28: $^{31}P\{^1H\}$ NMR (162 MHz, 298 K, C_6D_6 insert in dme) spectrum of the possible intermediate **22i** trapped in dme (dimethoxyethane). The intermediate slowly decomposes to complex **22**.

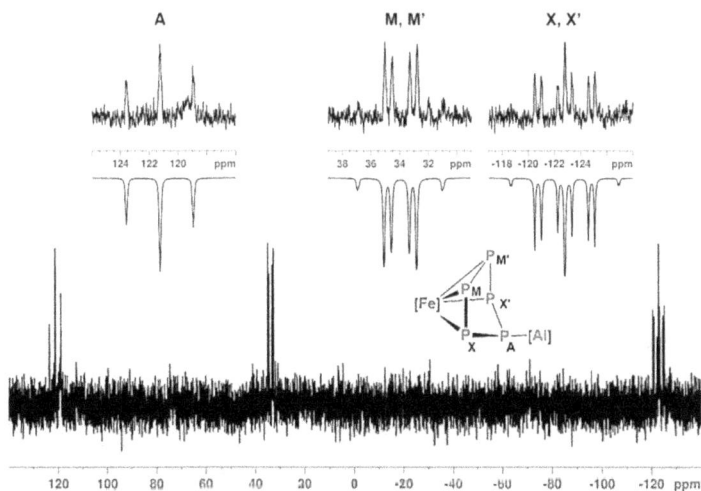

Figure 3.3.29: $^{31}P\{^1H\}$ NMR spectrum (162 MHz, 298 K, C_6D_6) of compound **22i** with nuclei assigned to an AMM'XX' spin system; insets: extended signals (upwards) and simulations (downwards); $\delta(P_X)$ = 121.3 ppm, $\delta(P_{MM'})$ = 33.9 ppm, $\delta(P_{AA'})$ = -122.8 ppm, $^1J_{MM'}$ = 378.2 Hz, $^1J_{AM}$ = $^1J_{A'M'}$ = 402.8 Hz, $^1J_{AX}$ = $^1J_{A'X}$ = -377.4Hz, $^2J_{AA'}$ = 10.8 Hz, $^2J_{A'M}$ = $^2J_{AM'}$ = -36.9 Hz, $^2J_{MX}$ = $^2J_{M'X}$ = -10.2 Hz, [Fe] = Cp*Fe, [Al] = Cp*Al(dme)$_n$.

Table 3.3.8: Chemical shifts, couplings constants and linewidths from the iterative fit of the AMM'XX' spin system of **22i** at 298 K and schematic representation of the FeP5-Al core. [Fe] = Cp*Fe, [Al] = Cp*Al(dme). Refinement with uncertainty of at least +/- 1 Hz because of poor resolution of the spectrum.

Parameters	Iteration values	Parameters	Iteration values
$\delta(P_X)$	121.25 ppm	$^1J_{MM'}$	378.2 Hz
$\delta(P_M) = \delta(P_{M'})$	33.91 ppm	$^1J_{AM} = {}^1J_{A'M'}$	402.8 Hz
$\delta(P_A) = \delta(P_{A'})$	-122.78 ppm	$^1J_{AX} = {}^1J_{A'X}$	-377.4 Hz
$\omega_{1/2}(X)$	13.8 Hz	$^2J_{AA'}$	10.8 Hz
$\omega_{1/2}(M) = \omega_{1/2}(M')$	16.2 Hz	$^2J_{A'M} = {}^2J_{AM'}$	-36.9 Hz
$\omega_{1/2}(A) = \omega_{1/2}(A')$	16.1 Hz	$^2J_{MX} = {}^2J_{M'X}$	-10.2 Hz

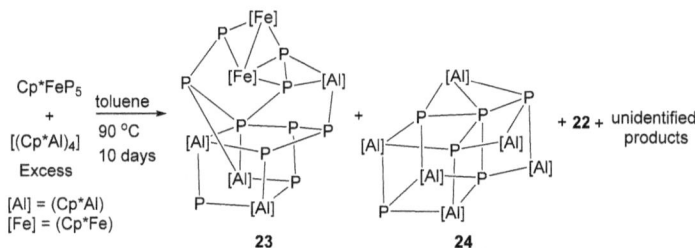

Scheme 3.3.14: Proposed structures of complexes **23** and **24** formed upon thermolysis of [Cp*Fe(η^5-P$_5$)] in the presence of excess [(Cp*Al)$_4$].

The insertion of three [Cp*Al^{III}]$^{2+}$ moieties into two P-P bonds of [Cp*Fe(η^5-P$_5$)], leading to the fragmentation of the *cyclo*-P$_5$ ring, motivated us to examine whether a further insertion of [Cp*Al^{III}]$^{2+}$ moieties would be possible. Heating [Cp*Fe(η^5-P$_5$)] in the presence of excess [(Cp*Al^{I})$_4$] at 90 °C for 10 days led to the formation of different products which could be successfully crystallized (Scheme 3.3.14). X-ray diffraction studies revealed the formation of **22** along with two new products, **23** and **24**. The solid-state structures of **23** and **24** showed the formation of an Al-Fe polyphosphide cluster and an aluminium polyphosphide cage complex, respectively (Figure 3.3.30). Unfortunately, complexes **23** and **24** were obtained in low yields and could not be isolated in a pure form due to similar solubilities and the formation of other unidentified products. The low yields of **23** and **24** prevented further analytical characterization, and establishment of their exact identity was also hindered by the ambiguous assignment of the Al and P atoms in the corresponding X-ray structures. By comparison of the bond lengths in **23** and **24** with previously reported Al-P, P-P, Al-C, and P-C bond lengths, a tentative assignment was nonetheless possible, and the suggested structures of complexes **23** and **24** are depicted in figure 3.3.30.

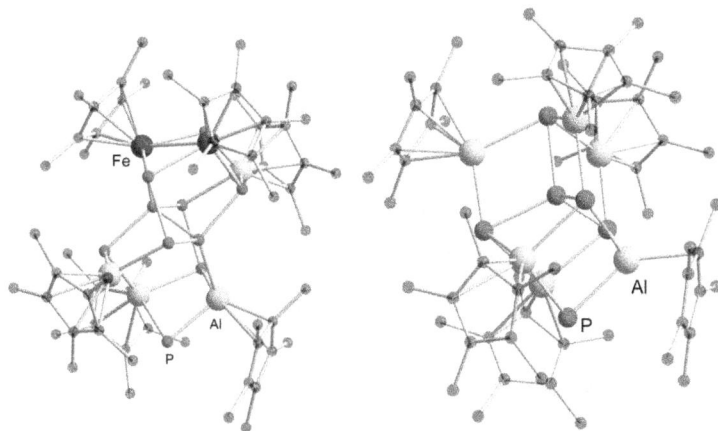

Figure 3.3.30: Possible molecular structures of **23** (left) and **24** (right). Colour code; Al (sky blue), P (purple), Fe (dark blue), and C (black). Hydrogen atoms are omitted for clarity.

4. Experimental Section

4.1 General Methods

All the manipulations of air- and water-sensitive reactions were performed with rigorous exclusion of oxygen and moisture in flame-dried Schlenk-type glassware either on a dual manifold Schlenk line, interfaced to a high vacuum (10^{-3} torr) line or in an argon-filled MBraun glove box. Hexane was distilled under nitrogen from potassium benzophenoneketyl before storage *in vacuo* over LiAlH$_4$. Toluene, diethylether, heptane, and pentane were dried by using an MBraun solvent purification system (SPS 800), degassed and stored *in vacuo* over LiAlH$_4$. Tetrahydrofuran and hexane were distilled under nitrogen from potassium benzophenoneketyl before storage over LiAlH$_4$. Dichloromethane was distilled under nitrogen from P$_2$O$_5$ and stored over 4 Å molecular sieves. Decalin was refluxed over sodium and distilled under vacuum before storing under nitrogen atmosphere. Deuterated solvents were obtained from Carl Roth (99 atom % D). CD$_2$Cl$_2$ and CDCl$_3$ were distilled from P$_2$O$_5$ and stored *in vacuo* over 4 Å molecular sieves. C$_6$D$_6$, d$_8$-toluene, and d$_8$- thf were degassed, dried, and stored *in vacuo* over Na/K alloy in resealable flasks. Elemental analyses were carried out with an Elementar vario Micro cube. IR spectra were obtained on a Bruker Tensor 37 spectrometer equipped with a room temperature DLaTGS detector and a diamond ATR (attenuated total reflection) unit. ^{1}H, ^{13}C{^{1}H}, ^{31}P{^{1}H}, and ^{29}Si{^{1}H}IG (inverse gated) NMR spectra were recorded on a Bruker Avance 400 (^{1}H: 400.30 MHz, ^{13}C: 100.67 MHz, ^{31}P: 162.04 MHz, ^{29}Si: 79.5 MHz) or on a Bruker Avance 300 (^{1}H: 300.13 MHz, ^{13}C: 75.48 MHz, ^{31}P: 121.50 MHz, ^{29}Si: 59.6 MHz). The chemical shifts are reported in ppm relative to external TMS (^{1}H, ^{13}C, ^{29}Si) and H$_3$PO$_4$ (85%) (^{31}P). Simulations of the ^{31}P{^{1}H} NMR spectra were performed using the DAISY module of the Topspin 3.6 processing software (Bruker). The parameters chemical shift (δ), coupling constant (J) and linewidth ($\omega_{1/2}$) for the simulation of the phosphorus NMR spectra are compiled in the different tables. The ratio between the different ^{29}Si substituted isomers was defined by the natural abundance of ^{29}Si (4.7%). The following reagents were purchased from commercial sources and used without further purification: 4-(diphenylphosphino)benzoic acid (H-**L^{Ph}**), 3-(diphenylphosphino)propionic acid (H-**L^{Et}**), Bipyridine, ZnMe$_2$, Cp*, [Fe(CO)$_5$], [Fe$_2$(CO)$_9$], [Fe$_3$(CO)$_{12}$], [Co$_2$(CO)$_8$], [Mn$_2$(CO)$_{10}$], [Re$_2$(CO)$_{10}$],

[('BuN)$_2$C], HSiCl$_3$, GeCl$_2$(1,4-dioxane). Note: NMR studies of complexes **5** to **12** were not possible due to paramagnetic character and low solubility in non-polar solvents.

4.2 Synthesis of Starting Materials

[(Bipy)ZnMe$_2$],[125] [AuCl(tht)],[128] [(DippForm)$_2$Ln^{II}(thf)$_2$] (Ln = Sm,[138] Yb[49]), [(Cp*)$_2$Sm^{II}(thf)$_2$],[33] [Cp*Fe(η^5-P$_5$)],[71] IPr,[172] ITMe,[197] LSiCl,[177b] LGeCl,[181] [LSi(N(SiMe$_3$)$_2$)],[180] [IPr-GeCl$_2$],[183] [LSi-SiL],[184] [LGe-GeL],[181] [Mes-BDI-Mg^I]$_2$,[193] [Dipp-BDI-Al^I],[117] [(Cp*Al^I)$_4$],[110] [Dipp-BDI-H],[198] [Mes-BDI-H],[198] [Ln^{II}I$_2$] (Ln = Sm, Yb)[199], and DippForm-H[200].

4.3 Synthesis and analytical data

4.3.1 Synthesis of [(Bipy)Zn(p-O$_2$C(C$_6$H$_4$)PPh$_2$)$_2$] (1)

Pre-cooled (*ca.* -30 °C) thf (15 mL) was added to a Schlenk flask containing a mixture of [(Bipy)ZnMe$_2$] (126 mg, 0.5 mmol) and H-L^{Ph} (306 mg, 1 mmol) *via* cannula. The color of the reaction mixture quickly changed from light yellow to colorless indicating the formation of complex **1**. The solution was allowed warm to room temperature. All the volatiles were removed after stirring the reaction mixture for 6 hours. The product was recrystallized by slow evaporation of a solution of **1** in dichloromethane and toluene (2:1). The colourless crystals were washed with 5 mL of ice-cold diethylether and 2*5 mL of pentane and dried *in vacuo*.

Yield = 70% (292 mg, 0.35 mmol). Anal. Calcd for C$_{48}$H$_{36}$N$_2$O$_4$P$_2$Zn (832.15): C, 69.28; H, 4.36; N, 3.37. Found: C, 69.93; H, 4.09; N, 3.67.

^{1}H NMR (400 MHz, 298 K, CDCl$_3$): δ [ppm] = 7.27-7.34 (m, 24 H, Ph-*H*), 7.62 (br, 2 H, NCHC*H*, $\Delta \nu_{1/2}$ ≈ 16 Hz), 8.02-8.06 (m, 6 H, NCHCHC*H* & Ph-*H*), 8.20 (br, 2 H, NCC*H*, $\Delta \nu_{1/2}$ ≈ 14 Hz), 9.13 (br, 2 H, NC*H*CH, $\Delta \nu_{1/2}$ ≈ 18 Hz).

^{31}P{^{1}H} NMR (162 MHz, 298 K, CDCl$_3$): δ [ppm] = -5.2 (s, *P*Ph$_2$Ar).

^{13}C{^{1}H} NMR (100 MHz, 298 K, CDCl$_3$): δ [ppm] = 121.2 (NC*C*H), 126.7 (N*C*HCH), 128.7 (d, *m-C*$_{Ph}$, $^3J_{CP}$ = 7 Hz), 129.0 (*p-C*$_{Ph}$), 130.20 (d, *m-C*$_{PhCOO}$, $^3J_{CP}$ = 7 Hz), 132.8 (*i-C*$_{PhCOO}$), 133.1 (d, *o-C*$_{PhCOO}$, $^2J_{CP}$ = 19.5 Hz), 134.0 (d, *o-C*$_{Ph}$, $^2J_{CP}$ = 20 Hz), 136.8 (d, *i-C*$_P$, $^1J_{CP}$ = 10.6 Hz), 140.5 (N*C*HCHCH), 142.1 (d, *i*-C*P*, $^1J_{CP}$ = 12.5 Hz), 150.3 (N*C*HCH), 174.5 (*C*OOR), No signal could be detected for N*C*CH of bipyridine ligand.

IR (ATR): $\tilde{v}$ (cm^{-1}) = 3052 (vw), 1615 (s), 1606 (s), 1594 (m), 1550 (m), 1535 (w), 1490 (m), 1473 (m), 1445 (s), 1435 (s), 1391 (s), 1391 (s), 1376 (s), 1361 (m), 1299 (w), 1250 (w), 1234 (m), 1200 (m), 1176 (w), 1155 (m), 1135 (w), 1088 (m), 1070 (m), 1058 (m), 1028 (m), 1015 (m), 999 (m), 983 (w), 912 (w), 855 (s), 846 (s), 812 (w), 779 (vs), 768 (vs), 752 (vs), 737 (vs), 723 (vs), 704 (vs), 692 (m), 650 (m), 633 (w), 586 (w), 554 (w), 542 (m), 526 (s), 504 (w), 481 (m), 433 (m), 413 (s).

4.3.2 Synthesis of [(Bipy)Zn(*p*-O$_2$C(C$_6$H$_4$)PPh$_2$(AuCl))$_2$](2a) and [(Bipy)$_2$Zn$_3${*p*-O$_2$C(C$_6$H$_4$)PPh$_2$(AuCl)}$_6$] (2)

To a mixture of **1** (100 mg, 0.12 mmol) and [AuCl(tht)] (77 mg, 0.24 mmol) was added thf (10 mL). The reaction mixture was stirred at room temperature for 3 hours. All the volatiles were removed *in vacuo* to obtain **2a**. The product was washed with 2*5 mL of pentane and dried under vacuum. A small amount of single crystals of only **2** were obtained from by evaporation of a solution of crude product in dichloromethane, acetone and ethanol in 1:1:1 ratio. The low solubility and low yield hampered further characterization of **2** in solution.

Analytical data for **2a**

Yield = 78% (121 mg, 0.09 mmol). Anal. Calcd for C$_{48}$H$_{36}$N$_2$O$_4$P$_2$ZnAu$_2$Cl$_2$ (1296.99): C, 44.45; H, 2.80; N, 2.16. Found: C, 43.95; H, 3.07; N, 1.83.

^{1}H NMR (400 MHz, 298 K, CDCl$_3$): δ [ppm] = 7.43-7.51 (m, 24 H, Ph-*H*), 7.69 (t, 2 H, NCHC*H*, $^3J_{HH}$ = 6 Hz), 8.06 (t, 2 H, NCHCHC*H*, $^3J_{HH}$ = 6 Hz), 8.19 (d, 4 H, Ph-*H*, $^3J_{HH}$ = 8.14 Hz), 8.27 (d, 2 H, NC*CH*, $^3J_{HH}$ = 8 Hz), 9.14 (d, 2 H, NC*H*CH, $^3J_{HH}$ = 4.5 Hz).

^{31}P{^{1}H} NMR (162 MHz, 298 K, CDCl$_3$): δ [ppm] = 33.4 (br, *P*-AuCl, $\Delta v_{1/2} \approx$ 120 Hz).

^{13}C{^{1}H} NMR (100 MHz, 298 K, CDCl$_3$): δ [ppm] = 121.4 (NC*C*H), 127.0 (N*C*HCH), 128.6 (d, *i*-C*P*, $^1J_{CP}$ = 30 Hz), 129.4 (d, *C*Ph, $^1J_{CP}$ = 6 Hz), 129.4 (d, *C*Ph, J_{CP} = 12 Hz), 131.0 (d, *C*Ph, J_{CP} = 12 Hz), 132.2 (*C*Ph), 133.7 (d, *C*Ph, J_{CP} = 14 Hz), 134.2 (d, *C*Ph, J_{CP} = 14 Hz), 137.3 (*C*Ph), 141.0 (NCHCH*C*H), 149.3 (N*C*CH), 150.3 (N*C*HCH), No signal could be detected for *C*OOR.

IR (ATR): $\tilde{v}$ (cm^{-1}) = 3056 (vw), 2945 (vw), 2883 (vw), 2839 (vw), 1599 (m), 1586 (m), 1538 (m), 1538 (m), 1493 (w), 1475 (m), 1436 (s), 1415 (s), 1389 (s), 1315 (w), 1306 (w), 1253 (w), 1186 (w), 1158 (w), 1101 (s), 1058 (w), 1027 (m), 1016 (m), 998 (m), 974 (w), 861 (w), 779 (s), 766 (s), 748 (vs), 732 (s), 713 (s), 700 (vs), 691 (vs), 654 (w), 633 (w), 618 (vw), 565 (m), 533 (s), 504 (s), 483 (w), 456 (s).

4.3.3 Synthesis of [(Bipy)Zn(O$_2$C(C$_2$H$_4$)PPh$_2$)$_2$] (3)

To a mixture of [(Bipy)ZnMe$_2$] (126 mg, 0.5 mmol) and H-L^{Et} (258 mg, 1 mmol) pre-cooled (*ca.* - 30 °C) thf (15 mL) was added. Likewise **1**, the reaction mixture quickly turned from light yellow to colorless. The reaction mixture was allowed to warm to room temperature and subsequently stirred for 6 hours. The colourless crystals of **3** were grown by vapor diffusion of pentane in thf solution. The mother liquor was separated by decantation and crystals were dried under vacuum.

Yield = 62% (250 mg, 0.34 mmol). Anal. Calcd for C$_{40}$H$_{36}$N$_2$O$_4$P$_2$Zn (736.07): C, 65.27; H, 4.93; N, 3.81. Found: C, 65.49; H, 4.73; N, 3.69.

^{1}H NMR (400 MHz, 298 K, CDCl$_3$): δ [ppm] = 2.32-2.43 (m, 8 H, CH$_2$CH$_2$), 7.25-7.27 (m, 12 H, Ph-*H*), 7.36-7.40 (m, 8 H, Ph-*H*), 7.60 (br, 2 H, NCHC*H*, $\Delta v_{1/2} \approx$ 20 Hz), 8.05 (t, 2 H, NCHCHC*H*, $^3J_{HH}$ = 7 Hz), 8.17 (d, 2 H, NCC*H*, $^3J_{HH}$ = 8 Hz), 8.98 (br, 2 H, NC*H*CH, $\Delta v_{1/2} \approx$ 13 Hz).

^{31}P{^{1}H} NMR (162 MHz, 298 K, CDCl$_3$): δ [ppm] = -14.7 (s, CH$_2$PPh$_2$).

^{13}C{^{1}H} NMR (100 MHz, 298 K, CDCl$_3$): δ [ppm] = 24.31 (d, CH$_2$PPh$_2$, $^1J_{CP}$ = 11 Hz), 31.55 (d, CH$_2$CO$_2$, $^2J_{CP}$ = 18 Hz), 121.0 (NCCH), 126.8 (NCHCH), 128.4 (d, *m*-C$_{Ph}$, $^3J_{CP}$ = 6.5 Hz), 128.5 (*p*-C$_{Ph}$), 133.9 (d, *o*-C$_{Ph}$, $^2J_{CP}$ = 18 Hz), 138.61 (d, *i*-CP, $^1J_{CP}$ = 12 Hz), 140.6 (NCHCHCH), 150.3 (NCHCH), 180.59 (d, CO$_2$R, $^3J_{CP}$ = 16 Hz), No signal could be detected for NCCH.

IR (ATR): $\tilde{v}$ (cm^{-1}) = 3071 (vw), 3019 (vw), 2952 (vw), 2908 (vw), 1708 (w), 1612 (s), 1600 (s), 1569 (m), 1476 (m), 1450 (m), 1434 (m), 1399 (s), 1319 (m), 1292 (s), 1270 (m), 1255 (m), 1195 (w), 1161 (m), 1099 (m), 1067 (w), 1043 (w), 1026 (m), 999 (w), 949 (w), 935 (w), 851 (vw), 826 (vw), 785 (vs), 753 (s), 742 (vs), 699 (vs), 681 (w), 659 (w), 637 (w), 610 (w), 527 (m), 513 (s), 475 (s), 432 (s).

4.3.4 Synthesis of [(Bipy)$_2$Zn$_3${O$_2$C(C$_2$H$_4$)PPh$_2$(AuCl)}$_6$] (4)

To a mixture of **3** (100 mg, 0.135 mmol) and [AuCl(tht)] (87 mg, 0.217 mmol) was added thf (10 mL). The reaction mixture was stirred for 3 hours at room temperature. All the volatiles were removed *in vacuo*. The residue was washed with 2*5 mL of pentane and dried under vacuum to obtained complex **4**. The product was dissolved in minimum amount of thf and allowed to stand in dark at room temperature. After a few weeks colourless single crystals suitable for X-ray diffraction were obtained.

Yield = 68% (104 mg, 0.015 mmol). Anal. Calcd for C$_{110}$H$_{100}$N$_4$O$_{12}$P$_6$Cl$_6$Au$_6$Zn$_3$ (3442.51): C, 38.33; H, 2.92; N, 1.63. Found: C, 38.45; H, 3.18; N, 1.60.

^{1}H NMR (400 MHz, 298 K, CDCl$_3$): δ [ppm] = 2.54-2.60 (m, 12 H, PCH_2), 2.73-2.79 (m, 12 H, CH_2CO$_2$), 7.41-7.54 (m, 36 H, Ph-*H*), 7.61-7.67 (m, 24 H, Ph-*H*), 7.69 (t, 4 H, NCHC*H*, $^3J_{HH}$ = 6.3 Hz), 8.14 (t, 4 H, NCHCHC*H*, $^3J_{HH}$ = 6.7 Hz), 8.23 (d, 4 H, NCC*H*, $^3J_{HH}$ = 8 Hz), 8.95 (d, 4 H, NC*H*CH, $^3J_{HH}$ = 4.8 Hz).

^{31}P{^{1}H} NMR (162 MHz, 298 K, CDCl$_3$): δ [ppm] = 29.5 (s, *P*-AuCl).

^{13}C{^{1}H} NMR (100 MHz, 298 K, CDCl$_3$): δ [ppm] = 24.4 (d, d, CH_2PPh$_2$, $^1J_{CP}$ = 39.66 Hz), 30.55 (CH_2CO$_2$), 121.2 (NCCH), 127.2 (N$CHCH$), 129.3 (*p*-C$_{Ph}$), 129.4 (d, *m*-C$_{Ph}$, $^3J_{CP}$= 11.5 Hz), 132.2 (d, *o*-C$_{Ph}$, $^2J_{CP}$ = 2.9 Hz), 133.36 (d, *i*-CP, $^1J_{CP}$ = 13 Hz), 141.1 (NCHCHCH), 149.0 (NCCH), 150.2 (NCHCH), 177.42 (d, CO$_2$R, $^3J_{CP}$ = 18 Hz).

IR (ATR): $\tilde{v}$ (cm^{-1}) = 3058 (vw), 3025 (vw), 2925 (vw), 2850 (vw), 1703 (w), 1598 (m), 1493 (w), 1474 (m), 1435 (m), 1376 (vw), 1314 (vw), 1265 (vw), 1184 (vw), 1157 (vw), 1104 (m), 1026 (vw), 998 (vw), 954 (vw), 901 (vw), 795 (vs), 769 (s), 738 (s), 693 (vs), 652 (w), 632 (w), 521 (s), 486 (s), 466 (w), 434 (w), 416 (w).

4.3.5 Synthesis of [{(DippForm)$_2$Sm^{III}}$_2${(μ_3-CO)$_2$(CO)$_9$Fe$_3$}] (5)

15 mL of thf was condensed at -78 °C onto a mixture of [(DippForm)$_2$Sm^{II}(thf)$_2$] (205 mg, 0.20 mmol) and [Fe$_3$(CO)$_{12}$] (51 mg, 0.10 mmol) or [Fe$_2$(CO)$_9$] (72 mg, 0.20 mmol) and warmed to room temperature and then stirred for 48 h at 60 °C. All the volatiles were removed *in vacuo*. Toluene

(15 mL) was added to the residue and refluxed for five minutes and filtered the hot reaction mixture. Red coloured crystals were obtained upon slowly cooling the filtrate to room temperature. The mother liquor was decanted off and the product was dried under vacuum.

Yield = 33% (74 mg, 0.033 mmol) (based on crystals). Anal. Calcd for $C_{111}H_{140}N_8O_{11}Fe_3Sm_2$ (2230.64): C, 59.77; H, 6.33; N, 5.02. Found: C, 59.77; H, 6.20; N, 4.81.

IR (ATR): $\bar{\nu}$ (cm^{-1}) = 2962 (vs), 2926 (m), 2869 (m), 2011 (vs), 1979 (vs), 1967 (s), 1878 (m), 1830 (w), 1696 (w), 1667 (s), 1640 (s), 1636 (m), 1586 (s), 1512 (s), 1464 (s) 1457 (m), 1437 (m), 1384 (s), 1362 (m), 1346 (m), 1332 (m), 1314 (s), 1278 (s), 1368 (m), 1254 (s), 1236 (m), 1186 (m), 1112 (w), 1098 (m), 1054 (m), 1042 (m), 1023 (w), 1004 (w), 947 (w), 935 (m), 823 (s), 800 (s) 768 (m), 753 (s), 681 (s), 645 (m), 612 (s), 585 (s), 508 (w), 474 (w), 458 (w), 444 (w), 438 (w), 433 (w), 420 (w).

4.3.6 Synthesis of [{(DippForm)$_2$Sm^{III}(thf)}$_2${(μ-CO)$_2$(CO)$_2$Co}$_2$] (6)

To a mixture of [(DippForm)$_2$Sm^{II}(thf)$_2$] (205 mg, 0.20 mmol) and [Co$_2$(CO)$_8$] (34 mg, 0.10 mmol) was condensed toluene (15 mL) at -78 °C and then stirred for 12 h at room temperature. The reaction mixture was filtered through P4 frit in a double ampule and flame sealed. Yellow crystals were grown by slow evaporation of toluene. Light yellow coloured crystals were washed carefully with toluene and dried under vacuum.

Yield = 56% (127 mg, 0.056 mmol) (based on crystals). Anal. Calcd for $C_{116}H_{156}N_8O_{10}Co_2Sm_2$ (2241.16): C, 62.17; H, 7.02; N, 5.00. Found: C, 62.34; H, 6.84; N, 4.81.

IR (ATR): $\bar{\nu}$ (cm^{-1}) = 2959 (s), 2927 (m), 2866 (m), 2020 (m), 1951 (br), 1935 (br), 1922 (br), 1904 (br), 1842 (s), 1819 (s), 1782 (s), 1665 (vs), 1636 (m), 1587 (m), 1527 (m), 1518 (m), 1464 (s), 1457 (s), 1439 (s), 1383 (s), 1361 (m), 1332 (m), 1314 (m), 1289 (m), 1272 (m), 1255 (m), 1236 (s), 1185 (m), 1107 (m), 1098 (m), 1057 (m), 1043 (m), 1016 (m), 934 (m), 912 (m), 865 (br), 800 (s), 753 (vs), 673 (s), 565 (w), 553 (s), 550 (s), 531 (s) 522 (s), 506 (s) 435 (w).

4.3.7 Synthesis of [{(DippForm)$_2$Yb^{III}(thf)}{(μ-CO)(CO)$_3$Co}] (7)

Following the procedure described above for **6**, the reaction of [(DippForm)$_2$Yb^{II}(thf)$_2$] (209 mg, 0.20 mmol) and [Co$_2$(CO)$_8$] (34 mg, 0.10 mmol) afforded orange crystals of **7**.

Yield = 63% (146 mg, 0.127 mmol) (based on crystals). Anal. Calcd for $C_{58}H_{78}N_4O_5CoYb$ (1143.27): C, 60.93; H, 6.88; N, 4.90. Found: C, 61.30; H, 6.89; N, 4.90.

IR (ATR): $\tilde{v}$ (cm^{-1}) = 2960 (s), 2927 (m), 2869 (m), 2031 (m), 2016 (m), 1915 (br), 1792 (m), 1748 (w), 1665 (vs), 1636 (m), 1587 (s), 1521 (m), 1465 (m), 1458 (m), 1439 (s), 1383 (s), 1361 (m), 1331 (m), 1331 (m), 1319 (m), 1290 (m), 1268 (m), 1255 (w), 1236 (m), 1186 (m), 1107 (m), 1107 (w), 1097 (w), 1058 (w), 1043 (w), 1025 (w), 1007 (w), 934 (w), 883 (w), 871 (w), 822 (w), 799 (s), 767 (w), 753 (w), 712 (w), 673 (w), 564 (w), 551 (vs), 510 (m), 434 (m), 418 (m).

4.3.8 Synthesis of [{(DippForm)$_2$Sm^{III}(thf)}{(μ-CO)(CO)$_4$Mn}] (8)

Toluene (15 mL) was condensed onto a mixture of [(DippForm)$_2$Sm^{II}(thf)$_2$] (205 mg, 0.20 mmol) and [Mn$_2$(CO)$_{10}$] (39 mg, 0.10 mmol) at -78 °C and the resulting solution was stirred for 16 h at 60 °C. The reaction mixture was filtered through a P4 frit into a double ampule and the latter was flame sealed. Yellow coloured crystals were grown by slow evaporation of toluene. The crystals were washed with a minimum amount of toluene and dried under vacuum.

Yield =56% (140 mg, 0.113 mmol) (based on crystals). Anal. Calcd for $C_{59}H_{78}N_4O_6MnSm*C_7H_8$ (8*toluene) (1236.73): C, 64.10; H, 7.01; N,4.53. Found: C, 64.29; H, 6.97; N, 4.50.

IR (ATR): $\tilde{v}$ (cm^{-1}) = 3061 (w), 2961 (s), 2927 (m), 2868 (m), 2061 (m), 2037(m), 2022 (m), 2012 (m), 1948 (m, sh), 1909 (s), 1734 (vw), 1665 (vs), 1588 (m), 1516 (w), 1456 (m), 1438 (m), 1383 (m), 1361 (m), 1315 (m), 1272 (s), 1255 (w), 1234 (m), 1190 (s), 1110 (m), 1098 (m), 1055 (w), 1044 (w), 1028 (w), 1004 (w), 985 (m), 945 (m), 933 (w), 831 (w), 800 (s), 770 (m), 755 (vs), 701 (vs), 680 (m), 664 (s), 641 (s), 571 (m), 555 (m), 517 (w), 462 (w), 426 (w), 414 (w).

4.3.9 Synthesis of [{(DippForm)$_2$Yb^{III}(thf)}{(μ-CO)(CO)$_4$Mn}] (9)

Following the procedure described above for **8**, the reaction of [(DippForm)$_2$Yb^{II}(thf)$_2$] (209 mg, 0.20 mmol) and [Mn$_2$(CO)$_{10}$] (39 mg, 0.10 mmol) afforded orange crystals of **9**.

Yield = 53% (130 mg, 0.107 mmol) (based on crystals). Anal. Calcd for $C_{59}H_{78}N_4O_6MnYb$ (1167.29): C, 60.71; H, 6.74; N, 4.80. Found: C, 60.71; H, 6.72, N, 4.85.

IR (ATR): $\tilde{v}$ (cm^{-1}) = 2961 (s), 2927 (m), 2865 (m), 2060 (w), 2038 (m), 2012 (m), 1951 (m), 1932 (s), 1906 (vs), 1701 (w), 1666 (vs), 1587 (m), 1521 (s), 1465 (s), 1457 (m), 1439 (s), 1385 (m), 1362 (m), 1332 (w), 1316 (m), 1289 (w), 1268 (s), 1255 (s), 1235 (m), 1193 (w), 1114 (w), 1098 (m), 1055 (w), 1044 (m), 1007 (w), 951 (w), 933 (w), 844 (w), 823 (m), 775 (w), 767 (m), 755 (vs), 724 (vs), 699 (w), 680 (m), 665 (m), 610 (w), 570 (m), 559 (m), 504 (w), 456 (w), 425 (m).

4.3.10 Synthesis of [{(DippForm)$_2$Sm^{III}(thf)}$_2${(μ-η^2-CO)$_2$(μ-η^1-CO)$_2$(CO)$_4$Re$_2$}] (10)

To a mixture of [(DippForm)$_2$Sm^{II}(thf)$_2$] (205 mg, 0.20 mmol) and [Re$_2$(CO)$_{10}$] (65 mg, 0.10 mmol) was condensed toluene (15 mL) at -78 °C and the resulting solution was stirred at 80 °C for 18 h. The reaction mixture was filtered through a P4 frit into a double ampule, and the latter was flame sealed. Red coloured crystals suitable for X-ray diffraction studies were obtained by slow evaporation of toluene. The crystals were washed with a minimum amount of toluene and dried under vacuum.

Yield = 62% (155 mg, 0.062 mmol) (based on crystals). Anal Calcd for C$_{116}$H$_{156}$N$_8$O$_{10}$Re$_2$Sm$_2$ (2495.70): C, 55.83; H, 6.30; N, 4.49. Found: C, 55.94; H, 5.76; N, 4.09.

IR (ATR): $\tilde{v}$ (cm^{-1}) = 2960 (vs), 2927 (m), 2866 (m), 2070 (w), 2012 (m), 1973 (s), 1903 (m), 1804 (w), 1664 (vs), 1636 (m), 1588 (m), 1520 (m), 1463 (m), 1438 (m), 1383 (m), 1361 (m), 1332 (m), 1317 (m), 1287 (m), 1271 (m), 1254 (m), 1235 (m), 1182 (m), 1110 (w), 1097 (w), 1057 (w), 1042 (vw), 1002 (w), 934 (w), 822 (w), 799 (w), 768 (m), 753 (m), 729 (m), 694 (m), 673 (w), 591 (m), 563 (w) 536 (w), 506 (vw), 464 (w), 435 (vw).

4.3.11 Synthesis of [{(Cp*)Sm$_2$^{III}}$_3${(Cp*)$_2$Sm^{III}(thf)}{(μ-O$_4$C$_4$)(μ-η^2-CO)$_2$(μ-η^1-CO)(CO)$_5$Re$_2$}] (11)

To a mixture of [(Cp*)$_2$Sm^{II}(thf)$_2$] (200 mg, 0.353 mmol) and [Re$_2$(CO)$_{10}$] (57 mg, 0.088 mmol) was condensed toluene (15 mL) at -78 °C, and the reaction mixture was stirred at room temperature for 48 h. A light-green coloured solid precipitated during the course of reaction. The reaction mixture was filtered through a P4 frit. The filtrate was concentrated to ca. 5 mL and allowed to stand at room temperature. Red coloured crystals were obtained after a few weeks. The mother liquor was decanted off, and the product was dried under vacuum.

Yield = 18% (40 mg, 0.016 mmol) (based on crystals). Anal. Calcd for $C_{96}H_{128}O_{13}Re_2Sm_4$ (2463.92): C, 46.80; H, 5.24. Found: C, 46.68; H, 5.57.

IR (ATR): $\tilde{v}$ (cm^{-1}) = 2962 (m), 2907 (m), 2854 (m), 2091 (vw), 2038 (w), 2010 (m), 1978 (vs), 1890 (m), 1858 (s), 1792 (s), 1777 (m), 1733 (s), 1663 (w), 1523 (m), 1437 (m), 1385 (w), 1377 (w), 1280 (w), 1253 (w), 1080 (vw), 1060 (vw), 1020 (w), 956 (w), 864 (vw), 837 (vw), 801 (vw), 784 (vw), 729 (m), 694 (w), 624 (w), 584 (s), 557 (m), 554 (vw), 463 (m), 435 (w) 419 (w).

4.3.12 Synthesis of [{(Cp*)$_2$Sm^{III}(thf)}{(μ-CO)$_2$(CO)$_3$Mn}]$_n$ (12)

Toluene (15 mL) was condensed onto a mixture of [(Cp*)$_2$Sm^{II}(thf)$_2$] (200 mg, 0.353 mmol) and [Mn$_2$(CO)$_{10}$] (68 mg, 0.176 mmol) at -78 °C and the reaction mixture was stirred at room temperature for 16 h. A red-coloured solid precipitated during the course of the reaction. All the volatiles were removed *in vacuo*. The residue was dissolved in hot thf (10 mL) and orange coloured crystals suitable for X-ray diffraction studies were obtained upon slowly cooling to room temperature. The mother liquor was decanted off and the product was dried under vacuum.

Yield = 62% (136 mg, 0.110 mmol) (based on crystals). Anal. Calcd for $C_{50}H_{60}O_{10}Mn_2Sm_2$ (**12**- thf) (1231.62): C, 48.76; H, 4.91. Found: C, 48.53 ; H 5.18.

IR (ATR): $\tilde{v}$ (cm^{-1}) = 2976 (m), 2911 (m), 2856 (m), 2013 (m), 1968 (s), 1940 (s), 1875 (w), 1830 (s), 1770 (vs), 1744 (s, sh), 1489 (vw), 1436 (m), 1388 (w), 1376 (w), 1246 (vw), 1062 (w), 1017 (w), 868 (br), 693 (vs), 684 (vs), 664 (vs), 588 (w), 541 (vw), 501 (w), 458 (w), 428 (vw).

4.3.13 Synthesis of complex [ITMe{(η^4-P$_5$)FeCp*}] (13)

To a mixture of [Cp*Fe(η^5-P$_5$)] (200 mg, 0.578 mmol) and ITMe (72 mg, 0.578 mmol), toluene (*ca.* 15 mL) was condensed at -78 °C, and the reaction mixture was allowed to warm up to room temperature. The reaction mixture was shortly heated to obtain a clear green-coloured solution and left undisturbed at room temperature. After 2 days, green coloured crystals were separated from the mother liquor by decantation. A second crop of crystals of **13** was obtained by storing the mother liquor at -30 °C for 12 h.

Combined yield of **13** = 92% (250 mg, 0.531 mmol). Anal Calcd for $C_{17}H_{27}N_2P_5Fe$ (470.13): C, 43.43; H, 5.79; N, 5.96. Found: C, 43.66; H, 5.47; N, 5.96.

^{1}H NMR (400 MHz, 298 K, C$_6$D$_6$): δ [ppm] = 0.77 (s, 6 H, C(CH_3)), 1.84 (s, 15 H, C(CH_3)), 2.86 (s, 6 H, N(CH_3))

^{13}C{^{1}H} NMR (100 MHz, 298 K, C$_6$D$_6$): δ [ppm] = 7.4 (C(CH$_3$)), 11.8 (C(CH$_3$)), 32.9 (C(CH$_3$)), 88.6 (C(CH$_3$)), 123.4 (N(CH$_3$)), NCN could not be detected.

^{31}P{^{1}H} NMR (162 MHz, <u>298 K</u>, C$_6$D$_6$): δ [ppm] = -47.5 (br, $\Delta \nu_{1/2} \approx$ 900 Hz), 37.9 (br, $\Delta \nu_{1/2} \approx$ 680 Hz).

^{31}P{^{1}H} NMR (162 MHz, <u>233 K</u>, thf -d_8): (AMM'XX') spin system, δ [ppm] = -51.0 (m, 2 P, P$_{XX'}$), 33.1 (m, 2 P, P$_{MM'}$), 39.2 (m, 1 P, P$_A$).

IR (ATR): $\tilde{\nu}$ (cm^{-1}) = 2964 (m), 2943 (m), 2898 (m), 2849 (m), 1647 (s), 1575 (m), 1476 (s), 1466 (s), 1433 (vs), 1392 (m), 1373 (vs), 1228 (m), 1169 (w), 1071 (m), 1056 (m), 1025 (vs), 845 (s), 797 (m), 753 (m), 666 (m), 617 (m), 589 (m), 564 (m), 494 (m), 468 (m), 468 (m), 444 (m), 430 (s).

4.3.14 Synthesis of complexes [(η^4-P$_4$SiL)FeCp*] (14) and [LSi(Cl)=P-SiL(Cl)$_2$] (15)

To a mixture of [Cp*Fe(η^5-P$_5$)] (195 mg, 0.564 mmol) and LSiCl (500 mg, 1.69 mmol), toluene (*ca.* 15 mL) was condensed at -78 °C and the reaction mixture was allowed to warm up to room temperature while stirring. After stirring at room temperature for 3 h, all the volatiles were removed *in vacuo*. The residue was extracted with *ca.* 15 mL of hexane and filtered. The hexane extract was concentrated until incipient crystallization and stored at -30 °C for 12 h to obtain yellow coloured crystals of **15**. The crystals were separated from the mother liquor and dried under vacuum. The remaining residue of the reaction mixture after hexane extraction was further extracted with *ca.* 30 mL of toluene and filtered. The filtrate was concentrated until incipient crystallization and stored at -30 °C for 12 h, leading to brown coloured crystals of **14**. The crystals were separated from the mother liquor and dried under vacuum. Single crystals of **14** suitable for X-ray crystallography were grown by slow evaporation of toluene.

Analytical data for complex **14**:

Yield of **14** = 40% (129 mg, 0.224 mmol). Anal. Calcd for C$_{25}$H$_{38}$N$_2$P$_4$SiFe (574.42): C, 52.27; H, 6.67; N, 4.88. Found: C, 52.36; H, 6.35; N, 4.88.

¹H NMR (400 MHz, 298 K, C₆D₆): δ [ppm] = 0.67 (s, 9 H, C(CH₃)₃), 1.27 (s, 9 H, C(CH₃)₃), 1.85 (s, 15 H, C(CH₃)), 6.50-6.85 (m, 5 H, C₆H₅).

¹³C{¹H} NMR (100 MHz, 298 K, C₆D₆): δ [ppm] = 11.8 (C(CH₃)), 31.7 (C(CH₃)₃), 32.4 (C(CH₃)₃), 54.94 (C(CH₃)₃), 54.76 (C(CH₃)₃), 87.6 (C(CH₃)), 127.7, 128.0, 129.9, 130.3 (C₆H₅), 172.3 (NCN).

³¹P{¹H} NMR (162 MHz, 298 K, C₆D₆): (AA'XX') spin system δ [ppm] = -194.4 (m, 2 P, P$_{XX'}$), 50.0 (m, 2 P, P$_{AA'}$).

²⁹Si{¹H} NMR (79.5 MHz, 298 K, C₆D₆): δ [ppm] = 42.3 (t, $^1J_{SiP}$ = 145 Hz).

IR (ATR): $\tilde{v}$ (cm⁻¹) = 2980 (m), 2965 (m), 2904 (m), 1602 (vw), 1519 (w), 1473 (m), 1453 (m), 1443 (m), 1406 (s), 1393 (s), 1375 (s), 1365 (s), 1278 (m), 1239 (m), 1226 (m), 1205 (m), 1180 (m), 1157 (m), 1089 (m), 1072 (m), 1025 (m), 931 (w), 893 (w), 796 (m), 770 (vs), 758 (vs), 728 (vs), 709 (m), 686 (s), 645 (vs), 588 (m), 560 (m), 496 (vs), 474 (m), 432 (w), 421 (w).

Analytical data for complex **15**:

Yield of **15** = 32% (120 mg, 0.182 mmol). Anal Calcd for C₃₀H₄₆N₄PCl₃Si₂ (656.22): C, 54.91; H, 7.07; N, 8.54. Found: C, 54.69; H, 6.80; N, 8.42.

¹H NMR (400 MHz, 298 K, C₆D₆): δ [ppm] = 1.47 (s, 36 H, C(CH₃)₃), 6.82-7.36 (m, 10 H, C₆H₅).

³¹P{¹H} NMR (162 MHz, 298 K, C₆D₆): δ [ppm] = -182.7 (s with ²⁹Si satellites, $^1J_{SiP}$ = 167 Hz).

¹³C{¹H} NMR (100 MHz, 298 K, C₆D₆): δ [ppm] = 31.8 (C(CH₃)₃), 56.1 (C(CH₃)₃), 128.5, 130.0 (Ph), the other signals of the Ph group could not be detected. Recording the ¹³C{¹H} NMR spectrum with large numbers of scans i.e. during longer measurement times led to low quality spectra because of decomposition products starting to appear, rendering the assignment of the carbons in the phenyl region vague.

²⁹Si{¹H} NMR: no signal could be obtained due to a low signal-to-noise ratio resulting from the coupling of the Si atoms with the P atom and decomposition of **15** in solution.

Experimental Section

IR (ATR): $\tilde{v}$ (cm^{-1}) = 2978 (m), 2965 (m), 2931 (m), 2908 (m), 2870 (w), 1616 (m), 1604 (m), 1548 (vw), 1520 (m), 1486 (m), 1475 (m), 1446 (m), 1402 (s), 1394 (vs), 1379 (vs), 1364 (vs), 1290 (w), 1229 (w), 1197 (s), 1180 (m), 1145 (m), 1100 (m), 1075 (m), 1044 (m), 1021 (m), 999 (m), 924 (w), 878 (m), 797 (s), 785 (s), 762 (m), 733 (m), 722 (s), 708 (s), 699 (s), 613 (s), 554 (vs), 547 (vs), 536 (vs), 491 (m), 465 (m), 437 (vw).

4.3.15 Synthesis of complex [{LSi(N(SiMe₃)₂)}{(η^4-P₅)FeCp*}] (16)

To a mixture of [Cp*Fe(η^5-P₅)] (82 mg, 0.237 mmol) and **L**Si(N(SiMe₃)₂) (100 mg, 0.237 mmol), toluene (*ca.* 15 mL) was condensed at -78 °C, and the reaction mixture was allowed to warm up to room temperature while stirring. After 6 h, the reaction mixture was filtered into a double ampoule and flame sealed under vacuum. Dark coloured crystals of **16** were grown by slow evaporation of toluene. The crystals of **16** were washed with 3 mL of cold toluene and dried under vacuum.

Yield of **16**·(0.5*toluene) = 79% (152 mg, 0.187 mmol). Anal Calcd for C₃₁H₅₆N₃P₅Si₃Fe·0.5(C₇H₈) (**16**·(0.5*toluene)) (811.85): C, 51.04; H, 7.45; N, 5.18. Found: C, 50.60; H, 7.47; N,4.99.

¹H NMR (400 MHz, 298 K, toluene-d_8): δ [ppm] = 0.13 (s, 9 H, Si(CH₃)₃), 0.64 (s, 9 H, Si(CH₃)₃), 1.12 (s, 18 H, C(CH₃)₃), 1.64 (s, 15 H, C(CH₃)), 6.83-6.98 (m, 4 H, C₆H₅), 8.67-8.70 (m, 1 H, C₆H₅).

¹³C{¹H} NMR (100 MHz, 298 K, toluene-d_8): δ [ppm] = 5.1 (Si(CH₃)₃), 6.5 (Si(CH₃)₃), 11.3 (C(CH₃)), 31.9 (C(CH₃)₃), 55.8 (C(CH₃)₃), 89.9 (C(CH₃)), 125.9, 127.6, 128.8, 129.3, 131.0, 132.4 (C₆H₅), 175.7 (NCN).

³¹P{¹H} NMR (162 MHz, 298 K, toluene-d_8): δ [ppm] = 35.3 (br, 5 P, $\Delta v_{1/2} \approx$ 1670 Hz).

³¹P{¹H} NMR (162 MHz, 193 K, toluene-d_8): (AA'MXX') spin system δ [ppm] = -29.9 (m, 2 P, P$_{XX'}$), 28.4 (m, 1 P, P$_M$), 30.7 (m, 2 P, P$_{AA'}$).

²⁹Si{¹H} NMR (79.5 MHz, 298 K, toluene-d_8): δ [ppm] = -34.9 (s, PSiN), 6.0 (s, NSi(CH₃)₃), 10.8 (s, NSi(CH₃)₃).

IR (ATR): $\tilde{v}$ (cm^{-1}) = 2972 (m), 2962 (m), 2902 (m), 1623 (w), 1598 (w), 1495 (w), 1475 (m), 1443 (m), 1371 (s), 1270 (s), 1254 (s), 1192 (s), 1094 (m), 1073 (m), 1028 (m), 1020 (m), 951 (vs), 912 (m), 862 (vs), 845 (m), 826 (m), 793 (m), 770 (m), 760 (m), 729 (m), 708 (m), 694 (m), 678 (m), 656 (m), 634 (m), 589 (w), 549 (w), 511 (m), 463 (m), 439 (m).

4.3.16 Synthesis of complex [{(η^4-P$_5$(SiL)$_2$}FeCp*] (17)

Toluene (15 mL) was condensed onto a mixture of [LSi-SiL] (100 mg, 0.192 mmol) and [Cp*Fe(η^5-P$_5$)] (67 mg, 0.192 mmol) at -78 °C, and the reaction mixture was allowed to warm to room temperature while stirring. After stirring at room temperature for 6 h, all the volatiles were removed *in vacuo*. The residue was extracted with *ca.* 15 mL of hexane and filtered. The hexane extract was concentrated until incipient crystallization and stored at -30 °C for 12 h to obtain yellow coloured crystals of **17**. The crystals were separated from the mother liquor and dried under vacuum. The remaining residue of the reaction mixture after the hexane extraction is a mixture of **17** (major) and **14** (minor).

Yield of **17** = 27% (45 mg, 0.052 mmol). Anal Calcd for C$_{40}$H$_{61}$N$_4$P$_5$Si$_2$Fe (864.84): C, 55.55; H, 7.11; N, 6.48. Found: C, 55.17; H, 6.87; N; 5.94.

^{1}H NMR (400 MHz, 298 K, C$_6$D$_6$): δ [ppm] = 1.46 (s, 18 H, C(CH$_3$)$_3$), 1.49 (s, 18 H, C(CH$_3$)$_3$), 1.96 (s, 15 H, C(CH$_3$)), 6.75-7.11 (m, 10 H, C$_6$H$_5$).

^{13}C{^{1}H} NMR (100 MHz, 298 K, C$_6$D$_6$): δ [ppm] = 11.2 (C(CH$_3$)), 32.0 (C(CH$_3$)$_3$), 32.1 (C(CH$_3$)$_3$), 54.9 (C(CH$_3$)$_3$), 55.8 (C(CH$_3$)$_3$), 90.6 (C(CH$_3$)), 128.82, 128.98, 129.9, 132.1 (C$_6$H$_5$), 170.7 (NCN).

^{31}P{^{1}H} NMR (162 MHz, 298 K, C$_6$D$_6$): (AA'MM'X) spin system, δ [ppm] = -163.5 (s, 1 P, P$_X$), -47.6 (m, 2 P, P$_{MM'}$), 80.8 (m, 2 P, P$_{AA'}$).

^{29}Si{^{1}H} NMR (79.5 MHz, 298 K, C$_6$D$_6$): δ [ppm] = 31.9 (dd, $^1J_{SiP}$ ≈ 165 and 118 Hz)

IR (ATR): $\tilde{v}$ (cm^{-1}) = 2971 (m), 2958 (m), 2927 (m), 2901 (m), 2869 (m), 1603 (vw), 1577 (vw), 1518 (m), 1472 (m), 1443 (m), 1410 (vs), 1392 (s), 1362 (s), 1274 (m), 1225 (m), 1206 (s), 1178 (m), 1088 (m), 1071 (m), 1043 (m), 1022 (m), 1022 (m), 927 (m), 893 (w), 851 (vw), 794 (m), 757 (vs), 728 (s), 707 (vs), 640 (m), 624 (vs), 616 (vs), 601 (vs), 568 (s), 507 (m), 496 (m), 475 (m), 441 (s).

4.3.17 Synthesis of [(LGe)$_2${(μ-η^4-P$_5$)FeCp*}] (18)

To a mixture of **LGe-GeL** (150 mg, 0.246 mmol) and [Cp*Fe(η^5-P$_5$)] (85 mg, 0.246 mmol) was condensed toluene (*ca.* 10 mL) at -78 °C and slowly warmed to room temperature. All the volatiles were removed under vacuum after stirring the reaction mixture for 30 minutes at room temperature. The residue was extracted in hexane (20 mL) and filtered. The filtrate was concentrated till incipient crystallization and subsequently stored at -30 °C for one day to obtain orange coloured crystals of complex [(**LGe**)$_2${(μ-η^4-P$_5$)FeCp*}]. The crystals were separated from mother liquor by decantation and dried *in vacuo*.

Yield of **18** = 60% (70 mg, 0.146 mmol). Anal Calcd. for C$_{40}$H$_{61}$N$_4$P$_5$Ge$_2$Fe (953.93): C, 50.36; H, 6.45; N, 5.87. Found: C, 49.62; H, 6.18; N, 5.60.

^{1}H NMR (400 MHz, 298 K, C$_6$D$_6$): δ [ppm] = 1.23 (s, 18 H, C(CH$_3$)$_3$), 1.32 (s, 18 H, C(CH$_3$)$_3$), 1.90 (s, 15 H, H-Cp*)), 6.89-7.02 (m, 6 H, *H*-Ph), 7.20 (t, 2 H, *H*-Ph, $^3J_{HH}$ = 7.2 Hz), 7.65 (d, 1 H, *H*-Ph, $^3J_{HH}$ = 6.5 Hz), 8.68 (d, 1 H, *H*-Ph, $^3J_{HH}$ = 7.2 Hz).

^{31}P{^{1}H} NMR (162 MHz, 298 K, C$_6$D$_6$): (AMM'XX') spin system δ [ppm] = -45.7 (m, 2 P, P$_{XX'}$), 43.5 (m, 2 P, P$_{MM}$), 150.5 (m, 1 P, P$_A$).

^{13}C{^{1}H} NMR (100 MHz, 298 K, C$_6$D$_6$): δ [ppm] = 11.9 (C(CH$_3$), Cp*), 32.3 (C(CH$_3$)$_3$), 32.6 (C(CH$_3$)$_3$), 53.5 (*C*(CH$_3$)$_3$), 54.4 (*C*(CH$_3$)$_3$), 89.8 (*C*(CH$_3$), Cp*), 127 (*C*-Ph), 129.6 (*C*-Ph), 129.7 (*C*-Ph), 131.2 (*C*-Ph), 135.5 (*C*-Ph), 136.35 (*C*-Ph), signals for NCN could not be detected.

IR (ATR): $\tilde{v}$ (cm^{-1}) = 2973 (s), 2960 (m), 2925 (m), 2900 (s), 2863 (s), 1646 (s), 1602 (s), 1578 (s), 1516 (s), 1474 (m), 1455 (m), 1412 (vs), 1389 (s), 1372 (s), 1359 (s), 1252 (s), 1221 (m), 1202 (s), 1177 (m), 1158 (m), 1140 (w), 1069 (m), 1063 (m), 1031 (m), 1019 (s), 925 (w), 898 (w), 849 (w), 791 (s), 741 (s), 706 (vs), 615 (w), 581 (m), 558 (w), 500 (w), 477 (m), 440 (m).

4.3.18 Isomerization of [(LGe)$_2${(μ-η^4-P$_5$)FeCp*}] (18) to [(LGe){(μ-η^3 P$_5$)(η^1-GeL)ΓeCp*}] (19) by 1,2-migration of germylene moiety

The isomerization of [(**LGe**)$_2${(μ-η^4-P$_5$)FeCp*}] to [(**LGe**){(μ-η^3-P$_5$)(η^1-GeL)FeCp*}] was monitored by ^{1}H and ^{31}P{^{1}H} NMR studies. In solution complex [(**LGe**)$_2${(μ-η^4-P$_5$)FeCp*}] slowly isomerizes to

[(LGe){(μ-η^3-P$_5$)(η^1-GeL)FeCp*}] at room temperature *via* a 1,2-migration of [LGe] moiety and subsequent coordination to [Cp*Fe]$^+$ moiety. This process slows down over the period and full conversion could not be obtained even after one month at room temperature. To increase the rate of reaction a solution of [(LGe)$_2${(μ-η^4-P$_5$)FeCp*}] in toluene was heated at 60 °C for 3 hours and only few yellowish-brown crystals of [(LGe){(μ-η^3-P$_5$)(η^1-GeL)FeCp*}] were obtained along with [Cp*Fe(η^5-P$_5$)] and other unidentified decomposed products as red-brown amorphous solid. Note: NMR spectrum of [(LGe){(μ-η^3-P$_5$)(η^1-GeL)FeCp*}] signals also have signals from complex [(LGe)$_2${(μ-η^4-P$_5$)FeCp*}] and [Cp*Fe(η^5-P$_5$)].

^{1}H NMR (400 MHz, 298 K, C$_6$D$_6$): δ [ppm] = 1.10 (s, 9 H, C(CH$_3$)$_3$), 1.31 (s, 9 H, C(CH$_3$)$_3$), 1.42 (s, 9 H, C(CH$_3$)$_3$), 1.52 (s, 9 H, C(CH$_3$)$_3$), 1.91 (s, 15 H, C(CH$_3$)), signals of phenyl group of [(LGe){(μ-η^1-η^3-P$_5$GeL)FeCp*}] could not be assigned due to overlapping signals of phenyl group of complex [(LGe)$_2${(μ-η^4-P$_5$)FeCp*}].

^{31}P{^{1}H} NMR (162 MHz, 298 K, C$_6$D$_6$): δ [ppm] = -43.0 (m, 1 P, P$_5$), -6.8 (m, 1 P, P$_4$), 72.4 (m, 1 P, P$_3$), 90.4 (m, 1 P, P$_2$), 252.5 (m, 1 P, P$_1$).

4.3.19 Synthesis of [(Mes-*BDI*-Mg^{II})$_2$(μ-η^4-η^4-P$_{10}$)(FeCp*)$_2$] (20)

To a mixture of [(Mes-*BDI*-Mg^I)$_2$] (100 mg, 0.139 mmol) and [Cp*Fe(η^5-P$_5$)] (96 mg, 0.279 mmol) was condensed toluene (*ca.* 10 mL) at -78 °C and slowly warmed to room temperature while stirring. The reaction mixture was further stirred for 3 hours and filtered. The filtrate was concentrated to *ca.* 5 mL under vacuum and stored at -30 °C. Red coloured crystals were obtained after few days. The crystals were separated from the mother liquor by decantation and dried *in vacuo*.

Yield of **20** = 52% (105 mg, 0.072 mmol). Anal Calcd. for C$_{66}$H$_{88}$N$_4$P$_{10}$Mg$_2$Fe$_2$*0.5(C$_7$H$_8$) (**20***(0.5 toluene)) (1453.56): C, 57.43; H, 6.38; N, 3.85. Found: C, 57.66; H, 5.96; N, 3.80.

^{1}H NMR (400 MHz, 298 K, C$_6$D$_6$): δ [ppm] = 1.21 (s, 30 H, C(CH$_3$)), 1.77 (s, 12 H, NCCH$_3$), 2.39 (s, 12 H, *p*-CH$_3$), 2.53 (s, 24 H, *o*-CH$_3$), 4.97 (s, 2 H, NCCHCN), 7.12 (s, 8 H, Ar-*H*).

^{31}P{^{1}H} NMR (162 MHz, 298 K, C$_7$D$_8$): δ [ppm] = -30 6 (br, $\Delta v_{1/2} \approx$ 600 Hz), 24.5 (br, $\Delta v_{1/2} \approx$ 550 Hz), 126.5 (br, $\Delta v_{1/2} \approx$ 500 Hz), 152.1 (br, $\Delta v_{1/2} \approx$ 500 Hz).

^{13}C{^{1}H} NMR (100 MHz, 298 K, C$_6$D$_6$): δ [ppm] = 11.2 (CCH$_3$), 21.4 (p-CH$_3$), 21.6 (o-CH$_3$), 24.1 (NCCH$_3$), 91.2 (CCH$_3$), 95.3 (NCCHCN), 130.4 (C-Ar), 132.7 (C-Ar), 133.9 (C-Ar), 145.9 (C-Ar), 169.34 (NCCH$_3$).

IR (ATR): $\tilde{v}$ (cm^{-1}) = 2972 (w), 2904 (m), 2853 (w), 1666 (vw), 1623 (w), 1609 (w), 1554 (m), 1517 (m), 1494 (m), 1477 (m), 1453 (s), 1391 (vs), 1372 (vs), 1302 (w), 1274 (m), 1258 (w), 1224 (s), 1197 (s), 1145 (s), 1070 (s), 1022 (w), 1008 (w), 957 (w), 924 (s), 853 (w), 828 (w), 797 (w), 741 (m), 728 (s), 694 (m), 651 (m), 627 (w), 598 (w), 565 (w), 537 (w), 502 (m), 465 (w), 437 (w), 427 (w).

4.3.20 Synthesis of [Dipp-BDI-Al^{III}(μ-η^4-P$_5$)FeCp*] (21)

A toluene (5 mL) solution of [Dipp-BDI-Al^I] (50 mg, 0.112 mmol) was added to a toluene (5 mL) solution of Cp*Fe(P$_5$) (39 mg, 0.112 mmol) at -78 °C and warmed to room temperature. The reaction mixture was stirred at room temperature for 1 hour. Then, it was transferred to double ampoule and flame sealed in vacuo. Yellow-green coloured single crystals were grown by slow evaporation of toluene. The mother liquor was decanted-off and crystals were dried under vacuum.

Yield of 21 = 35% (31 mg, 0.039 mmol) (based on crystals). Anal. Calcd for C$_{39}$H$_{56}$N$_2$P$_5$AlFe (790.59): C, 59.25; H, 7.14; N, 3.54. Found: C, 59.70; H, 7.38; N, 3.20.

^{1}H NMR (400 MHz, 298 K, C$_6$D$_6$): δ [ppm] = 1.04 (br, 12 H, CH(CH$_3$)$_2$, $\Delta v_{1/2} \approx$ 16 Hz), 1.22 (s, 15 H, C(CH$_3$)), 1.44 (s, 6 H, NC(CH$_3$)), 1.93 (d, 12 H, CH(CH$_3$)$_2$, $^3J_{HH}$ = 6.6 Hz), 3.37 (m, 4 H, CH(CH$_3$)), 4.71 (s, 1 H, NC(CH)CN), 7.29-7.35 (m, 6 H, Ph).

^{31}P{^{1}H} NMR (162 MHz, 298 K, C$_6$D$_6$): δ [ppm] = 32.0 (br, $\Delta v_{1/2} \approx$ 400 Hz), 60 7 (br, $\Delta v_{1/2} \approx$ 400 Hz), 98.4 (br, $\Delta v_{1/2} \approx$ 400 Hz).

^{31}P{^{1}H} NMR (162 MHz, 233 K, toluene-d$_8$): δ [ppm] = 30.7 (m, 2 P, P$_{XX'}$), 60.6 (m, 1 P, P$_M$), and 97.2 (m, 2 P, P$_{AA'}$).

^{13}C{^{1}H} NMR (100 MHz, 298 K, C$_6$D$_6$): δ [ppm] = 10.4 (C(CH$_3$)), 25.30 (CH$_3$CCHCCH$_3$), 25.34 (CH(CH$_3$)$_2$), 25.41 (CH(CH$_3$)$_2$), 29.27 (CH(CH$_3$)$_2$), 92.54 (C(CH$_3$)), 98 (CH$_3$CCHCCH$_3$), 125.1 (Ph), 128.57 (Ph), 128.70 (Ph), 129.33 (Ph), 141.6 (Ph), 172 (CN).

4.3.21 Synthesis of [(P)(Cp*Al^{III})$_3$(μ-η^2-η^2-η^2-η^4-P$_4$)(FeCp*)] (22)

15 mL toluene was condensed onto a mixture of [(Cp*Al)$_4$] (140 mg, 0.216 mmol) and [Cp*Fe(η^5-P$_5$)] (100 mg, 0.289 mmol). The reaction mixture was heated to 80 °C for one week. After cooling to room temperature the reaction mixture was filtered to a double ampoule and flame sealed *in vacuo*. Red coloured single crystals were grown by slow evaporation of toluene. The mother liquor was decanted-off and crystals were washed with cold toluene and dried under vacuum.

Yield of **22** = 47.6% (114 mg, 0.137mmol) (based on crystals). Anal. Calcd for C$_{40}$H$_{60}$P$_5$Al$_3$Fe (832.58): C, 57.71; H, 7.29. Found: C, 57.66; H, 7.23.

^{1}H NMR (400 MHz, 298 K, C$_6$D$_6$): δ [ppm] = 1.31 (s, 15 H, C(CH$_3$), [Cp*Fe]), 2.17 (s, 45 H, C(CH$_3$), [Cp*Al]).

^{31}P{^{1}H} NMR (162 MHz, 298 K, C$_6$D$_6$): δ [ppm] = -203.4 (s, 1 P), 73.8 (br, 4 P, $\Delta v_{1/2} \approx$ 110 Hz).

^{13}C{^{1}H} NMR (100 MHz, 298 K, C$_6$D$_6$): δ [ppm] = 10.41 (C(CH$_3$), [Cp*Fe]), 12.64 (C(CH$_3$), [Cp*Al]), 92.63 ((C(CH$_3$), [Cp*Fe]), 117.66 ((C(CH$_3$), [Cp*Al]).

4.3.22 Synthesis of [(Cp*Al^{III})$_4$(P$_{10}$)(Cp*Fe)$_2$] (23) and [(Cp*Al^{III})$_6$(P$_6$)] (24)

15 mL toluene was condensed onto a mixture of (Cp*Al)$_4$ (187 mg, 0.289 mmol) and [Cp*Fe(η^5-P$_5$)] (50 mg, 0.144 mmol). The reaction mixture was heated to 90 °C for 10 days. After cooling to room temperature the reaction was filtered to a double ampoule and flame sealed *in vacuo*. The mixture of crystals of **23** and **24** were grown by slow evaporation of toluene. Because of the low yields and similar solubility all the attempts to separate **23** and **24** were unsuccessful.

5. Crystal Structure Measurements

5.1 Data Collection and Refinement

A suitable crystal was covered in mineral oil (Aldrich) and mounted on a glass fibre. The crystal was transferred directly to the cold stream of a STOE IPDS 2 (150 or 210 K) or a STOE StadiVari (100 K) diffractometer. All structures were solved by using the program SHELXS/T[201] and Olex2.[202] The remaining non-hydrogen atoms were located from successive difference Fourier map calculations. The refinements were carried out by using full-matrix least-squares techniques on F^2 by using the program SHELXL.[201] In each case, the locations of the largest peaks in the final difference Fourier map calculations, as well as the magnitude of the residual electron densities, were of no chemical significance. The locations of the largest peaks in the final difference Fourier map calculation as well as the magnitude of the residual electron densities in each case were of no chemical significance. Atomic thermal displacement parameters for non-hydrogen atoms were refined anisotropically, with the exception of carbon atoms of severely disordered Cp* moiety in complex **18** and **20**.

In complex **2**, larger than expected residual density peaks around Au are due to truncation error and larger than expected residual density peaks around C32 and C50 are due to shadowing due to slight twinning.

Both solid state structure of **16** and **19** have two toluene molecules in the asymmetric unit cell and one toluene molecule could not be modelled satisfactorily and was therefore removed from the electron density map using the Olex2[202] solvent mask.

The crystal structure of **13** is twined. The twin was solved by using the following twin matrix: TWIN LAW (-1.0, 0.0, 0.0, 0.0, -1.0, 0.0, 0.0, 0.0, 1.0), BASF [0.417(3)].

5.2 Crystal Data

5.2.1 [(Bipy)Zn(p-O$_2$C(C$_6$H$_4$)PPh$_2$)$_2$] (1)

Compound	1
Formula	C$_{48}$H$_{36}$N$_2$O$_4$P$_2$Zn
$D_{calc.}$/ g cm^{-3}	1.379
μ/mm^{-1}	0.741
Formula Weight	832.10
Colour	colourless
Shape	block
T/K	150
Crystal System	triclinic
Space Group	$P\bar{1}$
a/Å	10.319(2)
b/Å	14.366(3)
c/Å	15.588(3)
α/°	110.17(3)
β/°	101.99(3)
γ/°	103.33(3)
V/Å^3	2003.6(9)
Z	2
Z'	1
Wavelength/Å	0.71073
Radiation type	MoK$_\alpha$
Θ_{min}/°	1.599
Θ_{max}/°	29.491
Measured Refl.	20827
Independent Refl.	11033
Reflections with I > 2(I)	8941
R_{int}	0.0160
Largest Peak	0.545
Deepest Hole	-0.323
GooF	1.049
wR_2 (all data)	0.0827
wR_2	0.0788
R_1 (all data)	0.0423
R_1	0.0305

5.2.2 [(Bipy)$_2$Zn$_3${p-O$_2$C(C$_6$H$_4$)PPh$_2$(AuCl)}$_6$] (2)

Compound	2
Formula	C$_{134}$H$_{100}$N$_4$O$_{12}$P$_6$Zn$_3$Au$_6$Cl$_6$
$D_{calc.}$/ g cm^{-3}	1.819
μ/mm^{-1}	7.190
Formula Weight	3734.60
Colour	colourless
Shape	prism
T/K	100
Crystal System	triclinic
Space Group	$P\bar{1}$
a/Å	11.160(2)
b/Å	17.074(3)
c/Å	19.451(4)
α/°	67.90(3)
β/°	84.83(3)
γ/°	83.93(3)
V/Å^3	3409.8(14)
Z	1
Z'	0.5
Wavelength/Å	0.71073
Radiation type	MoK$_\alpha$
Θ_{min}/°	2.264
Θ_{max}/°	29.511
Measured Refl.	31960
Independent Refl.	16308
Reflections with I > 2(I)	13606
R_{int}	0.0236
Largest Peak	15.328
Deepest Hole	-9.569
GooF	1.028
wR_2 (all data)	0.2491
wR_2	0.2380
R_1 (all data)	0.1032
R_1	0.0897

5.2.3 [(Bipy)Zn(O$_2$C(C$_2$H$_4$)PPh$_2$)$_2$] (3)

Compound	3
Formula	C$_{40}$H$_{36}$N$_2$O$_4$P$_2$Zn
$D_{calc.}$/ g cm^{-3}	1.359
μ/mm^{-1}	0.815
Formula Weight	736.02
Colour	colourless
Shape	plate
T/K	100
Crystal System	orthorhombic
Flack Parameter	0.037(8)
Hooft Parameter	0.011(7)
Space Group	$Aea2$
a/Å	26.1019(8)
b/Å	17.2340(4)
c/Å	7.9951(2)
α/°	
β/°	
γ/°	
V/Å^3	3596.52(16)
Z	4
Z'	0.5
Wavelength/Å	0.71073
Radiation type	MoK$_\alpha$
Θ_{min}/°	2.489
Θ_{max}/°	29.304
Measured Refl.	28358
Independent Refl.	4187
Reflections with I > 2(I)	3821
R_{int}	0.0321
Largest Peak	1.192
Deepest Hole	-0.389
GooF	1.038
wR_2 (all data)	0.0982
wR_2	0.0953
R_1 (all data)	0.0405
R_1	0.0360

5.2.4 [(Bipy)$_2$Zn$_3${O$_2$C(C$_2$H$_4$)PPh$_2$(AuCl)}$_6$] (4)

Compound	4*(4 thf)
Formula	C$_{126}$H$_{132}$N$_4$O$_{16}$P$_6$Zn$_3$Au$_6$Cl$_6$
$D_{calc.}$/ g cm^{-3}	1.921
μ/mm^{-1}	10.766
Formula Weight	3734.78
Colour	colourless
Shape	block
T/K	180
Crystal System	triclinic
Space Group	$P\bar{1}$
a/Å	13.2498(5)
b/Å	16.5272(5)
c/Å	16.8732(6)
α/°	106.106(3)
β/°	104.975(3)
γ/°	103.943(3)
V/Å^3	3228.9(2)
Z	1
Z'	0.5
Wavelength/Å	1.34143
Radiation type	GaK$_\alpha$
Θ_{min}/°	2.881
Θ_{max}/°	62.632
Measured Refl.	33754
Independent Refl.	15013
Reflections with I > 2(I)	12573
R_{int}	0.0293
Largest Peak	2.395
Deepest Hole	-1.462
GooF	1.087
wR_2 (all data)	0.1442
wR_2	0.1296
R_1 (all data)	0.0588
R_1	0.0477

5.2.5 [{(DippForm)$_2$Sm^{III}}$_2${(μ_3-CO)$_2$(CO)$_9$Fe$_3$}] (5)

Compound	5*(4 toluene)
Formula	C$_{139}$H$_{172}$N$_8$O$_{11}$Sm$_2$Fe$_3$
$D_{calc.}$/ g cm^{-3}	1.318
μ/mm^{-1}	1.267
Formula Weight	2599.09
Colour	clear red
Shape	block
T/K	100
Crystal System	monoclinic
Space Group	$P2_1/n$
a/Å	17.4276(4)
b/Å	20.8564(3)
c/Å	18.0530(4)
α/°	
β/°	93.692(2)
γ/°	
V/Å^3	6548.2(2)
Z	2
Z'	0.5
Wavelength/Å	0.71073
Radiation type	MoK$_\alpha$
Θ_{min}/°	1.853
Θ_{max}/°	28.614
Measured Refl.	67255
Independent Refl.	15423
Reflections with I > 2(I)	13087
R_{int}	0.0671
Largest Peak	1.675
Deepest Hole	-1.470
GooF	1.053
wR_2 (all data)	0.0971
wR_2	0.0942
R_1 (all data)	0.0418
R_1	0.0349

5.2.6 [{(DippForm)$_2$Sm^{III}(thf)}$_2${(μ-CO)$_2$(CO)$_2$Co}$_2$] (6)

Compound	6
Formula	C$_{116}$H$_{156}$N$_8$O$_{10}$Sm$_2$Co$_2$
$D_{calc.}$/ g cm^{-3}	1.361
μ/mm^{-1}	1.416
Formula Weight	2241.04
Colour	clear light yellow
Shape	plate
T/K	100
Crystal System	monoclinic
Space Group	$P2_1/n$
a/Å	11.942(2)
b/Å	24.391(5)
c/Å	18.910(4)
α/°	
β/°	96.75(3)
γ/°	
V/Å^3	5470.2(19)
Z	2
Z'	0.5
Wavelength/Å	0.71073
Radiation type	MoK$_\alpha$
Θ_{min}/°	2.094
Θ_{max}/°	26.000
Measured Refl.	26613
Independent Refl.	10715
Reflections with I > 2(I)	8775
R_{int}	0.0327
Largest Peak	1.364
Deepest Hole	-1.022
GooF	1.040
wR_2 (all data)	0.1008
wR_2	0.0927
R_1 (all data)	0.0533
R_1	0.0390

5.2.7 [{(DippForm)$_2$Yb^{III}(thf)}{(μ-CO)(CO)$_3$Co}] (7)

Compound	7
Formula	C$_{58}$H$_{78}$N$_4$O$_5$YbCo
$D_{calc.}$/ g cm^{-3}	1.345
μ/mm^{-1}	1.989
Formula Weight	1143.21
Colour	orange
Shape	fragment
T/K	150
Crystal System	triclinic
Space Group	$P\bar{1}$
a/Å	12.205(2)
b/Å	14.110(3)
c/Å	18.032(4)
α/°	82.87(3)
β/°	77.96(3)
γ/°	68.54(3)
V/Å^3	2822.5(12)
Z	2
Z'	1
Wavelength/Å	0.71073
Radiation type	MoK$_\alpha$
Θ_{min}/°	1.553
Θ_{max}/°	26.009
Measured Refl.	21947
Independent Refl.	11066
Reflections with I > 2(I)	9247
R_{int}	0.0260
Largest Peak	0.678
Deepest Hole	-0.398
GooF	0.968
wR_2 (all data)	0.0549
wR_2	0.0526
R_1 (all data)	0.0378
R_1	0.0258

5.2.8 [{(DippForm)$_2$Sm^{III}(thf)}{(μ-CO)(CO)$_4$Mn}] (8)

Compound	8* toluene
Formula	C$_{66}$H$_{86}$N$_4$O$_6$SmMn
$D_{calc.}$/ g cm^{-3}	1.298
μ/mm^{-1}	1.169
Formula Weight	1236.67
Colour	yellow
Shape	plate
T/K	150
Crystal System	triclinic
Space Group	$P\bar{1}$
a/Å	12.3657(3)
b/Å	14.0273(3)
c/Å	19.3568(5)
α/°	89.302(2)
β/°	74.549(2)
γ/°	78.249(2)
V/Å^3	3165.26(14)
Z	2
Z'	1
Wavelength/Å	0.71073
Radiation type	MoK$_\alpha$
Θ_{min}/°	1.793
Θ_{max}/°	29.264
Measured Refl.	31661
Independent Refl.	16880
Reflections with I > 2(I)	13537
R_{int}	0.0341
Largest Peak	0.805
Deepest Hole	-1.044
GooF	0.978
wR_2 (all data)	0.0840
wR_2	0.0802
R_1 (all data)	0.0489
R_1	0.0342

5.2.9 [{(DippForm)$_2$Yb^{III}(thf)}{(μ-CO)(CO)$_4$Mn}] (9)

Compound	9* 1.5 toluene
Formula	C$_{69.5}$H$_{90}$N$_4$O$_6$YbMn
$D_{calc.}$/ g cm^{-3}	1.357
μ/mm^{-1}	1.706
Formula Weight	1305.43
Colour	orange
Shape	fragment
T/K	100
Crystal System	triclinic
Space Group	$P\bar{1}$
a/Å	12.1458(6)
b/Å	13.5433(7)
c/Å	19.7641(10)
α/°	90.345(4)
β/°	96.876(4)
γ/°	98.184(4)
V/Å^3	3193.9(3)
Z	2
Z'	1
Wavelength/Å	0.71073
Radiation type	MoK$_\alpha$
Θ_{min}/°	2.431
Θ_{max}/°	31.064
Measured Refl.	30724
Independent Refl.	15702
Reflections with I > 2(I)	13513
R_{int}	0.0250
Largest Peak	0.891
Deepest Hole	-0.571
GooF	1.024
wR_2 (all data)	0.0697
wR_2	0.0665
R_1 (all data)	0.0400
R_1	0.0305

5.2.10 [{(DippForm)$_2$Sm^{III}(thf)}$_2${(μ-η^2-CO)$_2$(μ-η^1-CO)$_2$(CO)$_4$Re$_2$}] (10)

Compound	10* 2 toluene
Formula	C$_{130}$H$_{172}$N$_8$O$_{10}$Sm$_2$Re$_2$
$D_{calc.}$/ g cm^{-3}	1.445
μ/mm^{-1}	2.957
Formula Weight	1339.92
Colour	red
Shape	prism
T/K	150
Crystal System	monoclinic
Space Group	$P2_1/n$
a/Å	16.7981(4)
b/Å	19.3819(5)
c/Å	19.0314(6)
α/°	
β/°	96.237(2)
γ/°	
V/Å^3	6159.5(3)
Z	2
Z'	1
Wavelength/Å	0.71073
Radiation type	MoK$_\alpha$
Θ_{min}/°	1.504
Θ_{max}/°	29.544
Measured Refl.	32481
Independent Refl.	17078
Reflections with I > 2(I)	12990
R_{int}	0.0382
Largest Peak	1.769
Deepest Hole	-1.533
GooF	0.965
wR_2 (all data)	0.0836
wR_2	0.0786
R_1 (all data)	0.0512
R_1	0.0333

5.2.11 [{(Cp*)$_2$Sm^{III}}$_3${(Cp*)$_2$Sm^{III}(thf)}{(μ-O$_4$C$_4$)(μ-η^2-CO)$_2$(μ-η^1-CO)(CO)$_5$Re$_2$}] (11)

Compound	11* 4 toluene
Formula	C$_{124}$H$_{160}$O$_{13}$Sm$_4$Re$_2$
$D_{calc.}$/ g cm^{-3}	1.570
μ/mm^{-1}	3.998
Formula Weight	2832.31
Colour	red
Shape	prism
T/K	150(2)
Crystal System	monoclinic
Space Group	$P2_1/n$
a/Å	16.1898(3)
b/Å	26.5820(7)
c/Å	28.8793(6)
α/°	
β/°	105.441(2)
γ/°	
V/Å^3	11979.8(5)
Z	4
Z'	1
Wavelength/Å	0.71073
Radiation type	MoK$_\alpha$
Θ_{min}/°	1.315
Θ_{max}/°	25.600
Measured Refl.	73301
Independent Refl.	22410
Reflections with I > 2(I)	18613
R_{int}	0.0493
Largest Peak	3.323
Deepest Hole	-1.335
GooF	1.036
wR_2 (all data)	0.1234
wR_2	0.1133
R_1 (all data)	0.0581
R_1	0.0456

5.2.12 [{(Cp*)$_2$Sm^{III}(thf)}{(μ-CO)$_2$(CO)$_3$Mn}]$_n$ (12)

Compound	2(12* 1.5 thf)
Formula	C$_{128}$H$_{176}$O$_{27}$Sm$_4$Mn$_4$
$D_{calc.}$/ g cm^{-3}	1.505
μ/mm^{-1}	2.203
Formula Weight	2967.84
Colour	orange
Shape	rod
T/K	150
Crystal System	triclinic
Space Group	$P\bar{1}$
a/Å	12.6229(4)
b/Å	12.8766(4)
c/Å	20.6420(7)
α/°	101.808(3)
β/°	94.242(3)
γ/°	90.676(3)
V/Å^3	3273.91(19)
Z	1
Z'	0.5
Wavelength/Å	0.71073
Radiation type	MoK$_\alpha$
Θ_{min}/°	1.616
Θ_{max}/°	29.517
Measured Refl.	34794
Independent Refl.	18120
Reflections with I > 2(I)	12194
R_{int}	0.0591
Largest Peak	1.580
Deepest Hole	-1.262
GooF	0.953
wR_2 (all data)	0.1084
wR_2	0.0986
R_1 (all data)	0.0806
R_1	0.0446

5.2.13 [ITMe{(η^4-P$_5$)FeCp*}] (13)

Compound	13* benzene
Formula	$C_{23}H_{33}FeN_2P_5$
$D_{calc.}$/ g cm^{-3}	1.356
μ/mm^{-1}	0.874
Formula Weight	548.21
Colour	clear green
Shape	irregular
T/K	210
Crystal System	monoclinic
Space Group	$P2_1/n$
a/Å	9.9712(5)
b/Å	19.9914(14)
c/Å	13.4715(8)
α/°	
β/°	90.419(4)
γ/°	
V/Å^3	2685.3(3)
Z	4
Z'	1
Wavelength/Å	0.71073
Radiation type	MoK$_\alpha$
Θ_{min}/°	1.512
Θ_{max}/°	29.582
Measured Refl.	18967
Independent Refl.	7437
Reflections with I > 2(I)	4779
R_{int}	0.0550
Largest Peak	0.978
Deepest Hole	-1.278
GooF	1.104
wR_2 (all data)	0.2871
wR_2	0.2380
R_1 (all data)	0.1257
R_1	0.0835

5.2.14 [(η^4-P$_4$SiL)FeCp*] (14)

Compound	14
Formula	C$_{25}$H$_{38}$N$_2$P$_4$SiFe
$D_{calc.}$/ g cm^{-3}	1.319
μ/mm^{-1}	0.801
Formula Weight	574.39
Colour	clear brown
Shape	block
T/K	210
Crystal System	monoclinic
Space Group	$P2_1/c$
a/Å	12.9314(5)
b/Å	13.6586(4)
c/Å	17.2014(6)
α/°	
β/°	107.773(3)
γ/°	
V/Å^3	2893.19(18)
Z	4
Z'	1
Wavelength/Å	0.71073
Radiation type	MoK$_\alpha$
Θ_{min}/°	1.941
Θ_{max}/°	29.588
Measured Refl.	17001
Independent Refl.	7909
Reflections with I > 2(I)	4826
R_{int}	0.0354
Largest Peak	0.506
Deepest Hole	-0.318
GooF	1.031
wR_2 (all data)	0.1395
wR_2	0.1151
R_1 (all data)	0.0989
R_1	0.0491

5.2.15 [LSi(Cl)=P-SiL(Cl)₂] (15)

Compound	15
Formula	$C_{30}H_{46}N_4PSi_2Cl_3$
$D_{calc.}$/ g cm^{-3}	1.175
μ/mm^{-1}	0.379
Formula Weight	656.21
Colour	clear yellow
Shape	prism
T/K	210
Crystal System	triclinic
Space Group	$P\bar{1}$
a/Å	10.3097(4)
b/Å	18.2239(8)
c/Å	20.5037(8)
α/°	81.348(3)
β/°	78.086(3)
γ/°	82.795(3)
V/Å^3	3708.6(3)
Z	4
Z'	2
Wavelength/Å	0.71073
Radiation type	MoK$_\alpha$
Θ_{min}/°	1.428
Θ_{max}/°	26.838
Measured Refl.	30108
Independent Refl.	15616
Reflections with I > 2(I)	10756
R_{int}	0.0248
Largest Peak	0.495
Deepest Hole	-0.232
GooF	0.997
wR_2 (all data)	0.1181
wR_2	0.1085
R_1 (all data)	0.0660
R_1	0.0419

5.2.16 [{LSi(N(SiMe₃)₂)}{(η⁴-P₅)FeCp*}] (16)

Compound	16* toluene
Formula	$C_{38}H_{64}N_3P_5Si_3Fe$
$D_{calc.}$/ g cm⁻³	1.174
μ/mm⁻¹	0.607
Formula Weight	811.82
Colour	clear green
Shape	irregular
T/K	150
Crystal System	triclinic
Space Group	$P\bar{1}$
a/Å	14.9783(3)
b/Å	15.7492(3)
c/Å	22.0917(5)
α/°	78.249(2)
β/°	72.189(2)
γ/°	68.508(2)
V/Å³	4591.85(18)
Z	4
Z'	2
Wavelength/Å	0.71073
Radiation type	MoK_α
Θ_{min}/°	1.576
Θ_{max}/°	29.517
Measured Refl.	45473
Independent Refl.	25049
Reflections with I > 2(I)	19324
R_{int}	0.0216
Largest Peak	0.640
Deepest Hole	-0.504
GooF	1.024
wR_2 (all data)	0.0982
wR_2	0.0897
R_1 (all data)	0.0508
R_1	0.0351

5.2.17 [{(η^4-P$_5$(SiL)$_2$}FeCp*] (17)

Compound	17* toluene
Formula	$C_{47}H_{69}N_4P_5Si_2Fe$
$D_{calc.}$/ g cm^{-3}	1.229
μ/mm^{-1}	0.528
Formula Weight	956.94
Colour	clear yellow
Shape	irregular
T/K	100
Crystal System	monoclinic
Space Group	$P2_1/n$
a/Å	15.7545(3)
b/Å	13.7810(4)
c/Å	24.9938(5)
α/°	
β/°	107.695(2)
γ/°	
V/Å^3	5169.7(2)
Z	4
Z'	1
Wavelength/Å	0.71073
Radiation type	MoK$_\alpha$
Θ_{min}/°	2.354
Θ_{max}/°	29.500
Measured Refl.	27237
Independent Refl.	12808
Reflections with I > 2(I)	8625
R_{int}	0.0326
Largest Peak	0.406
Deepest Hole	-0.273
GooF	0.996
wR_2 (all data)	0.0941
wR_2	0.0810
R_1 (all data)	0.0845
R_1	0.0426

5.2.18 [(LGe)$_2${(μ-η^4-P$_5$)FeCp*}] (18)

Compound	18
Formula	C$_{40}$H$_{61}$N$_4$P$_5$Ge$_2$Fe
$D_{calc.}$/ g cm^{-3}	1.162
μ/mm^{-1}	1.532
Formula Weight	953.80
Colour	orange
Shape	plate
T/K	100
Crystal System	triclinic
Space Group	$P\bar{1}$
a/Å	12.773(3)
b/Å	13.408(3)
c/Å	17.265(4)
α/°	98.62(3)
β/°	110.39(3)
γ/°	92.41(3)
V/Å^3	2726.0(11)
Z	2
Z'	1
Wavelength/Å	0.71073
Radiation type	MoK$_\alpha$
Θ_{min}/°	1.728
Θ_{max}/°	31.541
Measured Refl.	26902
Independent Refl.	14039
Reflections with I > 2(I)	8228
R_{int}	0.0490
Largest Peak	0.772
Deepest Hole	-0.997
GooF	1.045
wR_2 (all data)	0.1904
wR_2	0.1674
R_1 (all data)	0.1163
R_1	0.0662

5.2.19 [(LGe){(μ-η³-P₅)(η¹-GeL)FeCp*}] (19)

Compound	19* toluene
Formula	$C_{47}H_{69}N_4P_5Ge_2Fe$
$D_{calc.}$/ g cm⁻³	1.238
μ/mm⁻¹	1.494
Formula Weight	1045.94
Colour	clear yellowish brown
Shape	plate
T/K	150
Crystal System	triclinic
Space Group	$P\bar{1}$
a/Å	12.132(2)
b/Å	14.161(3)
c/Å	16.767(3)
α/°	84.70(3)
β/°	79.64(3)
γ/°	83.19(3)
V/Å³	2806.2(10)
Z	2
Z'	1
Wavelength/Å	0.71073
Radiation type	MoK_α
Θ_{min}/°	1.716
Θ_{max}/°	29.484
Measured Refl.	28121
Independent Refl.	15293
Reflections with I > 2(I)	9970
R_{int}	0.0389
Largest Peak	0.739
Deepest Hole	-0.378
GooF	1.007
wR_2 (all data)	0.0917
wR_2	0.0826
R_1 (all data)	0.0791
R_1	0.0408

5.2.20 [(Mes-*BDI*-Mg^{II})$_2$(μ-η^4-η^4-P$_{10}$)(FeCp*)$_2$] (20)

Compound	20* toluene
Formula	C$_{73}$H$_{96}$N$_4$P$_{10}$Mg$_2$Fe$_2$
$D_{calc.}$/ g cm^{-3}	1.274
μ/mm^{-1}	0.634
Formula Weight	1499.55
Colour	green
Shape	fragments
T/K	150
Crystal System	orthorhombic
Space Group	*Pbcn*
a/Å	22.400(5)
b/Å	28.977(6)
c/Å	24.096(5)
α/°	
β/°	
γ/°	
V/Å^3	15641(5)
Z	8
Z'	1
Wavelength/Å	0.71073
Radiation type	MoK$_\alpha$
Θ_{min}/°	1.149
Θ_{max}/°	25.578
Measured Refl.	48841
Independent Refl.	14634
Reflections with I > 2(I)	9137
R_{int}	0.0574
Largest Peak	1.229
Deepest Hole	-0.592
GooF	1.038
wR_2 (all data)	0.1920
wR_2	0.1592
R_1 (all data)	0.1280
R_1	0.0695

5.2.21 [Dipp-*BDI*-Al^{III}(μ-η^4-P$_5$)FeCp*] (21)

Compound	21
Formula	C$_{39}$H$_{56}$N$_2$P$_5$AlFe
$D_{calc.}$/ g cm^{-3}	1.295
μ/mm^{-1}	0.621
Formula Weight	790.53
Colour	clear greenish yellow
Shape	plate
T/K	100
Crystal System	monoclinic
Space Group	$P2_1/n$
a/Å	13.1200(9)
b/Å	23.212(2)
c/Å	13.6497(11)
$\alpha/°$	
$\beta/°$	102.676(6)
$\gamma/°$	
V/Å^3	4055.6(6)
Z	4
Z'	1
Wavelength/Å	0.71073
Radiation type	MoK$_\alpha$
$\Theta_{min}/°$	1.763
$\Theta_{max}/°$	30.440
Measured Refl.	25656
Independent Refl.	10637
Reflections with I > 2(I)	6295
R_{int}	0.0671
Largest Peak	0.615
Deepest Hole	-1.010
GooF	1.023
wR_2 (all data)	0.1895
wR_2	0.1543
R_1 (all data)	0.1305
R_1	0.0661

Crystal Structure Measurements

5.2.22 [(P)(Cp*Al^{III})$_3$(μ-η^2-η^2-η^2-η^4-P$_4$)(FeCp*)] (22)

Compound	22
Formula	C$_{40}$H$_{60}$P$_5$Al$_3$Fe
$D_{calc.}$/ g cm^{-3}	1.291
μ/mm^{-1}	0.628
Formula Weight	832.52
Colour	clear red
Shape	plate
T/K	100
Crystal System	triclinic
Space Group	$P\bar{1}$
a/Å	11.3626(12)
b/Å	11.7476(12)
c/Å	16.3303(17)
α/°	84.556(9)
β/°	88.702(9)
γ/°	80.771(9)
V/Å^3	2141.8(4)
Z	2
Z'	1
Wavelength/Å	0.71073
Radiation type	MoK$_\alpha$
Θ_{min}/°	1.764
Θ_{max}/°	31.824
Measured Refl.	20959
Independent Refl.	11529
Reflections with I > 2(I)	6147
R_{int}	0.0778
Largest Peak	1.146
Deepest Hole	-0.882
GooF	1.017
wR_2 (all data)	0.2333
wR_2	0.1858
R_1 (all data)	0.1653
R_1	0.0791

5.2.23 [(Cp*Al^{III})$_4$(P$_{10}$)(Cp*Fe)$_2$] (23)

Compound	23* 0.5(toluene)
Formula	C$_{63.5}$H$_{94}$P$_{10}$Al$_4$Fe$_2$
$D_{calc.}$/ g cm^{-3}	1.206
μ/mm^{-1}	0.670
Formula Weight	1386.70
Colour	clear dark brown
Shape	plate
T/K	150
Crystal System	triclinic
Space Group	$P\bar{1}$
a/Å	12.2828(7)
b/Å	13.5483(8)
c/Å	25.2149(15)
α/°	90.655(5)
β/°	99.853(5)
γ/°	112.099(5)
V/Å^3	3817.2(4)
Z	2
Z'	1
Wavelength/Å	0.71073
Radiation type	MoK$_\alpha$
Θ_{min}/°	1.628
Θ_{max}/°	29.550
Measured Refl.	38900
Independent Refl.	20897
Reflections with I > 2(I)	11311
R_{int}	0.0542
Largest Peak	3.323
Deepest Hole	-1.005
GooF	1.049
wR_2 (all data)	0.2888
wR_2	0.2616
R_1 (all data)	0.1455
R_1	0.0913

5.2.24 [(Cp*Al^{III})$_6$(P$_6$)] (24)

Compound	24* 2(toluene)
Formula	C$_{74}$H$_{106}$Al$_6$P$_6$
$D_{calc.}$/ g cm^{-3}	1.206
μ/mm^{-1}	0.257
Formula Weight	1343.28
Colour	clear yellow
Shape	plate
T/K	150
Crystal System	orthorhombic
Space Group	$Pnma$
a/Å	31.534(2)
b/Å	17.1803(18)
c/Å	13.6542(9)
α/°	
β/°	
γ/°	
V/Å^3	7397.3(10)
Z	4
Z'	0.5
Wavelength/Å	0.71073
Radiation type	MoK$_\alpha$
Θ_{min}/°	1.625
Θ_{max}/°	25.599
Measured Refl.	16606
Independent Refl.	7064
Reflections with I > 2(I)	3615
R_{int}	0.1573
Largest Peak	0.932
Deepest Hole	-0.817
GooF	1.040
wR_2 (all data)	0.3478
wR_2	0.2798
R_1 (all data)	0.1944
R_1	0.1106

6. Summary (Zusammenfassung)

6.1 Summary

The work reported in this thesis is divided into three parts. In the first part, the synthesis of Zn-Au heterometallic complexes is described by using bifunctional carboxy-phosphine ligands. The zinc-metalloligand **1**, bearing a para-substituted phenylene spacer, was synthesized by reaction of [(Bipy)ZnMe$_2$] with H-**L**^{Ph} (Scheme 6.1). Further reaction of complex **1** with [AuCl(tht)] resulted in the formation of a trinuclear Zn-Au complex (**2a**). Upon crystallization, a nonanuclear heterometallic complex (**2**) was obtained *via* the loss of one bipyridine ligand in every three units of **2a** (Scheme 6.1). In the solid state, complex **2** does not feature any inter- or intramolecular aurophilic interactions, possibly due to the rigidity of the phenylene spacer. A more flexible zinc-metalloligand (**3**), bearing an alkyl spacer (**L**^{Et}), was synthesized using a similar procedure as for **1**. Complex **3** furnished a nonanuclear Zn-Au heterometallic complex **4** upon reaction with

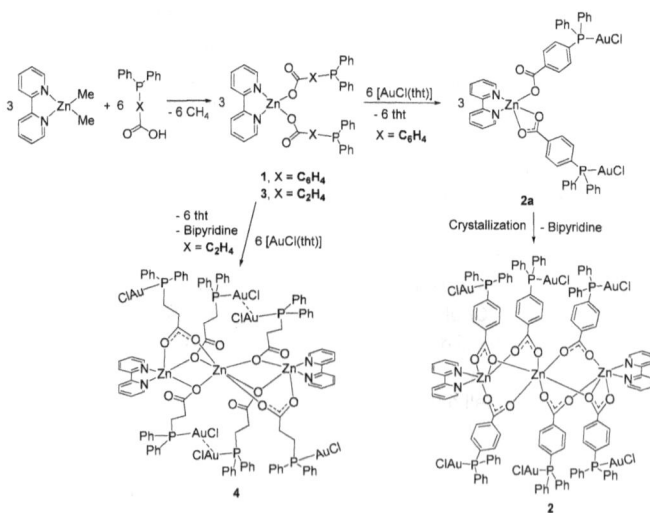

Scheme 6.1: Synthesis of zinc-metalloligands (**1** and **3**) as well as their corresponding Zn-Au heterometallic complexes (**2a**, **2**, and **4**).

[AuCl(tht)]. The latter exhibits intramolecular aurophilic interactions. The presence of aurophilic interactions in complex **4** can be mainly attributed to the flexibility of the L^{Et} ligand as compared to the more rigid L^{Ph} ligand.

The second part of this thesis focuses on redox reactions between divalent lanthanides (Ln^{II}) and transition metal (TM) carbonyl complexes for the synthesis of high nuclearity heterometallic complexes. The reaction between $[(DippForm)_2Sm^{II}(thf)_2]$ (DippForm = N,N'-bis(2,6-diisopropylphenyl)formamidinate) and $[Fe_2(CO)_9]$ or $[Fe_3(CO)_{12}]$ allowed the isolation of a pentanuclear complex, $[\{(DippForm)_2Sm^{III}\}_2\{(\mu_3\text{-}CO)_2(CO)_9Fe_3\}]$ (**5**). Whereas in case of $[Co_2(CO)_8]$ and $[(DippForm)_2Ln^{II}(thf)_2]$ (Ln = Sm, Yb) a tetranuclear complex $[\{(DippForm)_2Sm^{III}(thf)\}_2\{(\mu\text{-}CO)_2(CO)_2Co\}_2]$ (**6**) and a dinuclear complex $[\{(DippForm)_2Yb^{III}(thf)\}\{(\mu\text{-}CO)(CO)_3Co\}]$ (**7**) were obtained. The formation of both complexes is accompanied by reductive homolytic Co-Co bond cleavage.

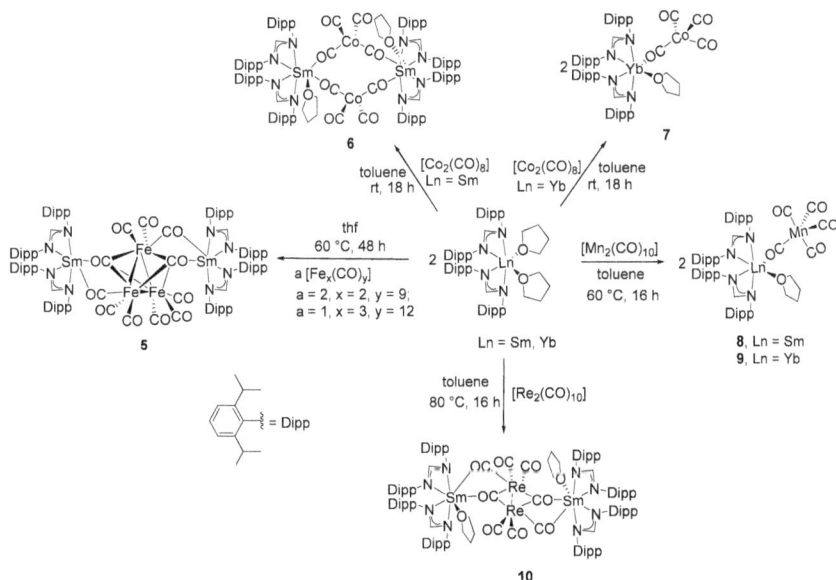

Scheme 6.2: Reactivity of TM carbonyl complexes with bulky divalent lanthanide complexes.

Similar homolytic Mn-Mn bond cleavage was observed in case of reaction between [(DippForm)$_2$Ln^{II}(thf)$_2$] (Ln = Sm, Yb) and [Mn$_2$(CO)$_{10}$] resulting in the isostructural products [{(DippForm)$_2$Ln^{II}(thf)}{(μ-CO)(CO)$_4$Mn}] (Ln = Sm (8), Yb (9)). In contrast to [Mn$_2$(CO)$_{10}$], [Re$_2$(CO)$_{10}$] furnished complex 10 upon reaction with [(DippForm)$_2$Sm^{II}(thf)$_2$], featuring a novel rhenium carbonyl anion, [(μ-η^2-CO)$_2$(μ-η^1-CO)$_2$(CO)$_4$Re$_2$]$^{2-}$, located in the coordination sphere of two [(DippForm)$_2$Sm^{III}(thf)]$^+$ moieties (Scheme 6.2).

Owing to the unusual reactivity observed in the case of [Re$_2$(CO)$_{10}$], the effect of the nature of the ligand on the reaction products was also studied. The reaction of [Re$_2$(CO)$_{10}$] with [(Cp*)$_2$Sm^{II}(thf)$_2$] (Cp* = C$_5$Me$_5$) resulted in the formation of an unprecedented hexanuclear cluster, [{(Cp*)$_2$Sm^{III}}$_3${(μ-O$_4$C$_4$)(μ-η^2-CO)$_2$(μ-η^1-CO)(CO)$_5$Re$_2$}Sm^{III}(Cp*)$_2$(thf)] (11) containing a five-membered rhenacycle. The formation can be explained by a tetramerization of the CO ligands through reductive C-C coupling reactions. In contrast, a 1D coordination polymer, [{(Cp*)$_2$Sm^{III}(thf)}{(μ-CO)$_2$(CO)$_3$Mn}]$_n$ (12), was obtained from reaction between [(Cp*)$_2$Sm^{II}(thf)$_2$] and [Mn$_2$(CO)$_{10}$]. These results highlight that the outcomes of the reactions between Ln^{II} complexes and TM carbonyls are greatly influenced by different factors: the nature of the Ln^{II} metal centre (Yb vs. Sm), the supporting ligands, as well as the TM carbonyl.

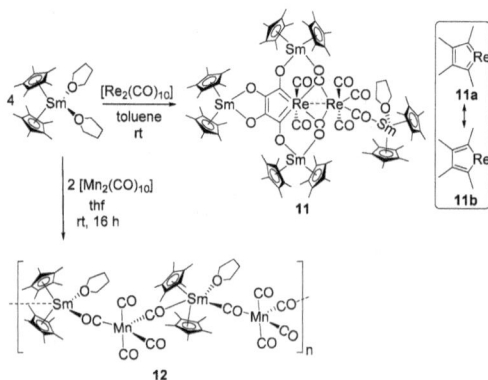

Scheme 6.3: Reactivity of samarocene with [Mn$_2$(CO)$_{10}$] and [Re$_2$(CO)$_{10}$] and possible double Fischer-carbene (11a) and metallacyclopentadiene (11b) forms.

The third and final part of this thesis focuses on reactivity studies of [Cp*Fe(η^5-P$_5$)] with low-valent main group compounds. Although no reaction was observed with the bulky IPr (IPr = 1,3-bis(2,6-diisopropylphenyl)imidazol-2-ylidene) N-heterocyclic carbene (NHC), even at elevated temperature (80 °C), using the more nucleophilic carbene, ITMe (ITMe = 1,3,4,5-tetramethylimidazol-2-ylidene), led to the formation of [ITMe{(η^4-P$_5$)FeCp*}] (13) (Scheme 6.4). The subsequent reaction of [Cp*Fe(η^5-P$_5$)] with the chloro-silylene [LSiCl] (L = PhC(N^{t}Bu)$_2$) resulted in the formation of [{(η^4-P$_4$SiL)FeCp*}] (14) via substitution of one phosphorous atom by an [LSi] moiety in the cyclo-P$_5$ ring. The eliminated P atom is scavenged by two [LSiCl] molecules to form LSi(Cl)=P-SiL(Cl)$_2$ (15) as a by-product. Complex [{LSi(N(SiMe$_3$)$_2$)}{(η^4-P$_5$)FeCp*}] (16) was obtained by using of a slightly modified silylene, [LSi(N(SiMe$_3$)$_2$)], with [Cp*Fe(η^5-P$_5$)]. Surprisingly, [Cp*Fe(η^5-P$_5$)] did not react with chloro-germylenes, such as [GeCl$_2$(1,4-dioxane)], [IPr-GeCl$_2$], or [LGeCl], even at high temperature. The reaction between [Cp*Fe(η^5-P$_5$)] and the di-silylene, [LSi-SiL], allowed the isolation of [{(η^4-P$_5$(SiL)$_2$}FeCp*}] (17) along with 14 and [L$_2$Si$_2$P$_2$]. Complex 17 consists of a seven-membered Si$_2$P$_5$ ring coordinated to the [Cp*Fe]$^+$ moiety. The seven-membered Si-P ring is formed by the reductive insertion of two [LSi]$^{3+}$ moieties into two adjacent P-P bonds of the cyclo-P$_5$ ring. Additionally, the reactivity of [Cp*Fe(η^5-P$_5$)] was studied towards the analogous di-germylene, [LGe-GeL], and a bis-germylene in coordination of polyphosphide was isolated as complex [(LGe)$_2${(μ-η^4-P$_5$)FeCp*}] (18). Unlike 17, complex 18

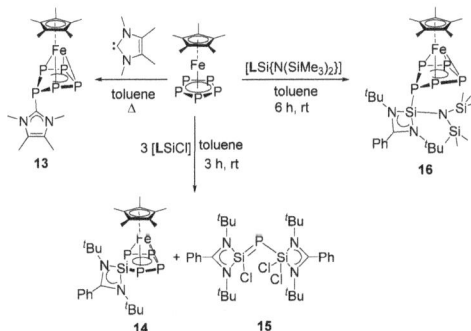

Scheme 6.4: Reactivity of [Cp*Fe(η^5-P$_5$)] with carbenes and silylenes.

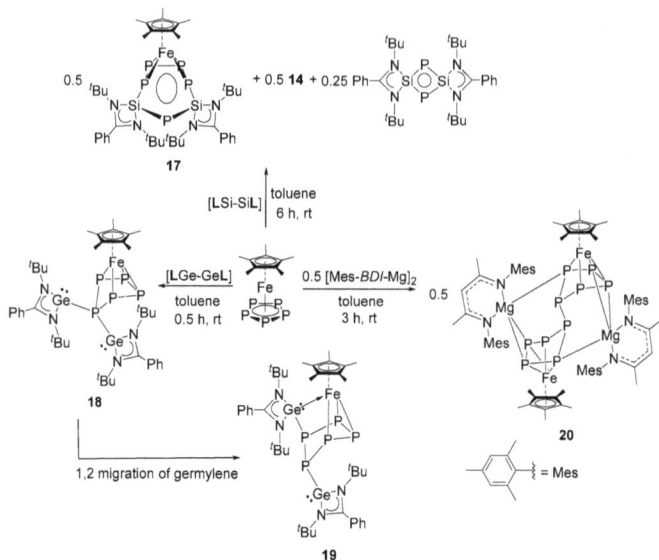

Scheme 6.5: Reactivity of [Cp*Fe(η^5-P$_5$)] with mono-valent Si, Ge, and Mg compounds.

features an intact *cyclo*-P$_5$ ring. NMR studies showed that complex **18** slowly converts to another product in solution at room temperature. X-ray analysis of the corresponding thermolysis product revealed formation of [(LGe){(μ-η^3-P$_5$)(η^1-LGe)FeCp*}] **(19)** *via* 1,2 migration of one [LGe]$^+$ moiety on the *cyclo*-P$_5$ ring. The migrated [LGe]$^+$ moiety further coordinates to the Fe centre in replacement of one P ligand. Also the largest magnesium polyphosphide complex [(Mes-*BDI*-Mg^{II})$_2$(μ-η^4-η^4-P$_{10}$)(FeCp*)$_2$] **(20)** was isolated by the reduction of [Cp*Fe(η^5-P$_5$)] with monovalent magnesium [Mes-*BDI*-Mg^I]$_2$ (Mes-*BDI* = (2,4,6-Me$_3$C$_6$H$_3$NCMe)$_2$CH) (Scheme 6.5).

The reactivity of [Cp*Fe(η^5-P$_5$)] was further expanded to monovalent aluminium complexes. The reaction between [Dipp-*BDI*-Al^I] (Dipp-*BDI* = (2,6-^{i}Pr$_2$C$_6$H$_3$NCMe)$_2$CH) and [Cp*Fe(η^5-P$_5$)] resulted in the formation of a neutral heterometallic Al-Fe triple-decker complex [(Dipp-*BDI*-Al^{III})(μ-η^3-η^4-P$_5$)FeCp*] **(21)** (Scheme 6.6). The reaction of the tetrameric monovalent [(Cp*Al^I)$_4$] with [Cp*Fe(η^5-P$_5$)] at 80 °C resulted in a tetrametallic Al-Fe polyphosphide cluster, [(P)(Cp*Al^{III})$_3$(μ-η^2-η^2-η^2-η^4-P$_4$)(FeCp*)] **(22)** (Scheme 6.6). Upon insertion of the [Cp*Al^{III}]$^{2+}$ moieties, the *cyclo*-P$_5$ ring is cleaved into an *acyclic*-P$_4$ chain and a P unit. Under harsher reaction conditions, the cyclo-P_5 ring undergoes complete disassembly resulting in Al-Fe polyphosphide **23** and Al polyphosphide **24** clusters.

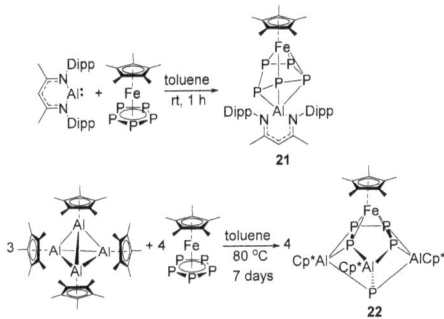

Scheme 6.6: Synthesis of aluminium polyphosphides.

6.2 Zusammenfassung

Diese Arbeit ist in drei Abschnitte unterteilt. Im ersten Abschnitt werden die Synthesen mehrkerniger heterometallischer Komplexe mittels bifunktioneller Carboxy-Phosphinliganden beschrieben. Der Zink-Metalloligand **1**, welcher einen para-substituierten Phenylen-spacer besitzt, wurde durch Umsetzung von [(Bipy)ZnMe₂] mit H-L^{Ph} synthetisiert (Schema 6.7). Die Reaktion von **1** mit [AuCl(tht)] resultierte in dem dreikernigen Zn-Au Komplex (**2a**). Des Weiteren konnte bei der Kristallisation die Ausbildung eines neunkernigen heterometallischen Komplexes (**2**) beobachtet werden, welcher durch die Abspaltung eines Bipyridin-Liganden von **2a** entsteht (Schema 6.7). Für Komplex **2** werden im Festkörper keine aurophilen Wechselwirkungen beobachtet, was vermutlich durch die Starrheit des Phenylen-Spacers verursacht wird. Ein flexiblerer Zink-Metalloligand (**3**), mit einem Alkyl-Spacer (**L^{Et}**), wurde in ähnlicher Weise wie

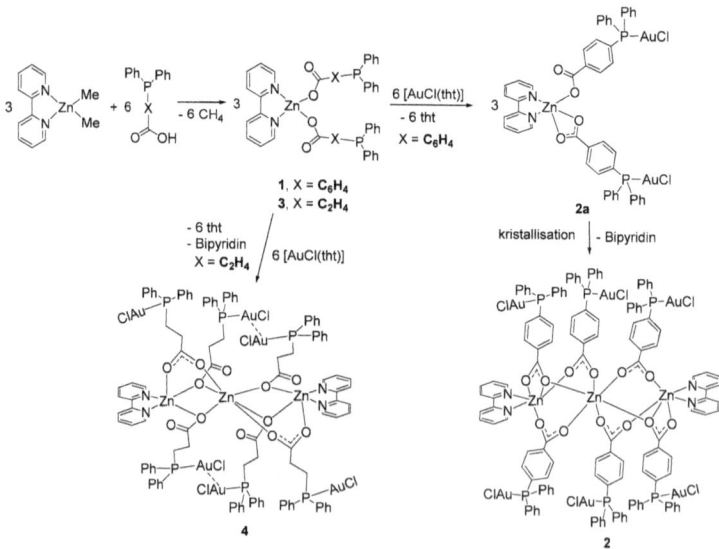

Schema **6.7**: Synthese der Zink-Metalloliganden **1** und **3** sowie der entsprechenden Zn-Au heterometallischen Komplexe (**2a**, **2** und **4**).

Verbindung **1** dargestellt. Die Reaktion von **3** mit[AuCl(tht)] resultierte in dem neunkernigen Zn-Au Komplex **4**. Dieser weist intramolekulare aurophile Wechselwirkungen auf, Diese werden durch die Flexibilität des **L^{Et}** Liganden ermöglicht.

Der zweite Teil dieser Arbeit behandelt die Synthese heterobimetallische Komplexe durch Redoxreaktionen zwischen zweiwertigen Lanthanoid- und Übergangsmetall-Carbonyl-Komplexen. Die Reaktion von [(DippForm)$_2$Sm^{II}(thf)$_2$] (DippForm = *N,N'*-Bis(2,6-diisopropylphenyl)formamidinat) mit [Fe$_2$(CO)$_9$] oder [Fe$_3$(CO)$_{12}$] führte zur Ausbildung des fünfkernigen Metallkomplexes [{(DippForm)$_2$Sm^{III}}$_2${(μ_3-CO)$_2$(CO)$_9$Fe$_3$}] (**5**). Hingegen kann bei einer Reaktion von [(DippForm)$_2$Ln^{II}(thf)$_2$] (Ln = Sm, Yb) mit [Co$_2$(CO)$_8$] die Bildung des vierkernigen Metallkomplexes [{(DippForm)$_2$Sm^{III}(thf)}$_2${(μ-CO)$_2$(CO)$_2$Co}$_2$] (**6**) und des zweikernigen Metallkomplexes [{(DippForm)$_2$Yb^{III}(thf)}{(μ-CO)(CO)$_3$Co}] (**7**) beobachtet werden.

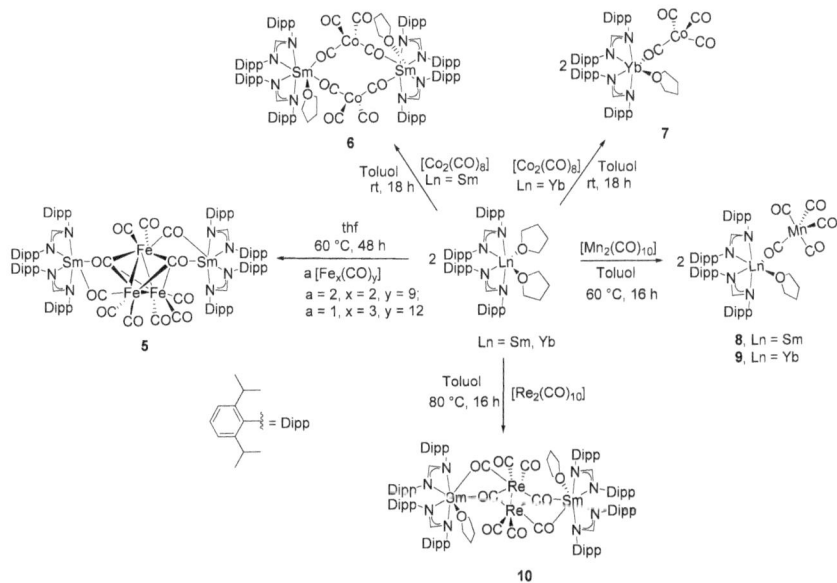

Schema 6.8: Reaktivität von Übergansmetall-Carbonylkomplexen mit sterisch anspruchsvollen zweiwertigen Lanthanoidkomplexen.

Die Synthese beider Komplexe geht mit einer reduktiven homolytischen Spaltung der Co-Co Bindung einher. Die Reduktion von $[Mn_2(CO)_{10}]$ mit $[(DippForm)_2Ln^{II}(thf)_2]$ (Ln = Sm, Yb) führte zu den isostrukturellen Produkten, $[\{(DippForm)_2Ln^{II}(thf)\}\{(\mu\text{-}CO)(CO)_4Mn\}]$ (Ln = Sm (8), Yb (9)). Diese werden durch eine homolytische Spaltung der Mn-Mn Bindung, über einen SET von den zweiwertigen Lanthanoidkomplexen zum Metall-Carbonyl Fragment, gebildet. Die Reaktion von $[(DippForm)_2Sm^{II}(thf)_2]$ mit $[Re_2(CO)_{10}]$ führte zur Bildung von Komplex 10. Dieser enthält ein neuartiges Rheniumcarbonylanion $[(\mu\text{-}\eta^2\text{-}CO)_2(\mu\text{-}\eta^1\text{-}CO)_2(CO)_4Re_2]^{2-}$, welches durch die Koordination zweier $[(DippForm)_2Sm^{III}(thf)]^+$-Einheiten stabilisiert wird (Schema 6.8). Aufgrund der beobachteten ungewöhnlichen Reaktivität von $[Re_2(CO)_{10}]$, wurde der Einfluss des Liganden auf die Reaktionsprodukte untersucht. Die Umsetzung von $[Re_2(CO)_{10}]$ mit $[(Cp^*)_2Sm^{II}(thf)_2]$ (Cp* = C_5Me_5) führte zur Ausbildung des sechskerniger Clusters $[\{(Cp^*)_2Sm^{III}\}_3\{(\mu\text{-}O_4C_4)(\mu\text{-}\eta^2\text{-}CO)_2(\mu\text{-}\eta^1\text{-}CO)(CO)_5Re_2\}Sm^{III}(Cp^*)_2(thf)]$ (11). Dieser Cluster besteht aus einen fünfgliedrigen Rhenacyclus, dessen Bildung durch die Tetramerisierung der CO-Liganden mittels reduktiver C-C Kupplung erklärt werden kann. Im Unterschied hierzu führte die Reaktion von $[(Cp^*)_2Sm^{II}(thf)_2]$ und $[Mn_2(CO)_{10}]$ zu einem 1D-Koordinationspolymer $[\{(Cp^*)_2Sm^{III}(thf)\}\{(\mu\text{-}CO)_2(CO)_3Mn\}]_n$ (12), welches durch eine Mn-Mn Bindungsspaltung gebildet wird. Diese Ergebnisse zeigen, dass

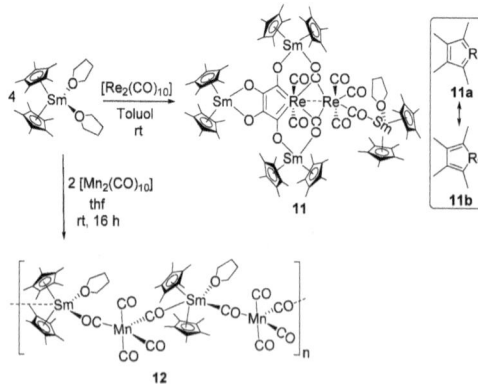

Schema 6.9: Reaktivtät von Samarocen mit $[Mn_2(CO)_{10}]$ und $[Re_2(CO)_{10}]$; sowie mögliche doppelte Fischer-Carben-Form (11a) und Metallacyclopentadien-Form (11b).

Reaktionen zwischen zweiwertigen Lanthanoidkomplexen und Übergangsmetall-Carbonylen im Wesentlichen durch drei verschiedene Faktoren beeinflusst werden: die Art des Lanthanoids (Yb vs. Sm), die unterstützenden Liganden sowie die verwendeten Übergangsmetall-Carbonyle.

Der dritte und letzte Teil dieser Arbeit behandelt die Reaktionen von [Cp*Fe(η^5-P$_5$)] mit niedervalenten Hauptgruppenverbindungen. Die Reaktion mit einem N-Heterocyclischen Carben (NHC) ITMe (ITMe = 1,3,4,5-tetramethylimidazol-2-yliden) führte zur Ausbildung des Komplexes [ITMe{(η^4-P$_5$)FeCp*}] (13) (Schema 6.10). Weiterhin ergab die Reaktion von [Cp*Fe(η^5-P$_5$)] mit dem Chloro-Silylen [LSiCl] (L = PhC(N^{t}Bu)$_2$) den Komplex [(η^4-P$_4$SiL)FeCp*] (14), welcher durch Substitution eines P-Atoms durch eine [LSi]-Einheit im cyclo-P$_5$ Ring des [Cp*Fe(η^5-P$_5$)] entsteht. Das substituierte P-Atom wird durch in der Lösung vorhandene [LSiCl] Moleküle abgefangen, wobei es zur Ausbildung von LSi(Cl)=P-SiL(Cl)$_2$ (15) als Nebenprodukt kommt. Es zeigte sich, dass letztere Verbindung bei Raumtemperatur nicht stabil ist und sich langsam weiter zu den bekannten Verbindungen [LSiCl$_3$] und [L$_2$Si$_2$P$_2$] zersetzt. Komplex [{LSi(N(SiMe$_3$)$_2$)}{(η^4-P$_5$)FeCp*}] (16) wurde aus der Reaktion des leicht modifizierten Silylens [LSi(N(SiMe$_3$)$_2$)] mit [Cp*Fe(η^5-P$_5$) isoliert. Interessanterweise reagiert [Cp*Fe(η^5-P$_5$)], auch bei erhöhten Temperaturen nicht mit sterisch anspruchsvollen NHCs (z. B. .1,3-Bis(2,6-diisopropylphenyl)imidazol-2-yliden)) und Chloro-Germylenen, wie z.B. dem [GeCl$_2$(1,4-dioxan)], [IPr-GeCl$_2$], oder [LGeCl] Die Reaktion von

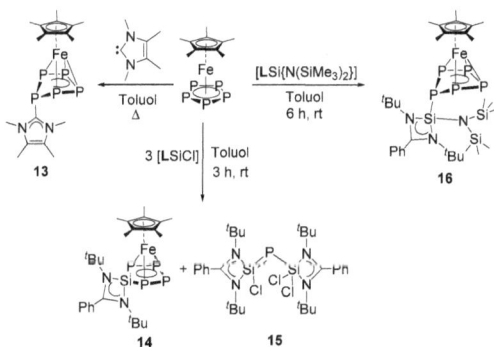

Schema 6.10: Reaktivität von [Cp*Fe(η^5-P$_5$)] mit Carbenen und Silylenen.

143

[Cp*Fe(η^5-P$_5$)] mit dem Di-Silylen [LSi-SiL], resultierte in den Verbindungen [{(η^4-P$_5$(SiL)$_2$}FeCp*] (17), Komplex 14 und [L$_2$Si$_2$P$_2$]. Komplex 17 besteht aus einem siebengliedrigen Si$_2$P$_5$-Ring, welcher an die [Cp*Fe]$^+$-Einheit koordiniert ist. Der siebengliedrige Si-P-Ring wird durch eine reduktive Insertion zweier [LSi]$^+$-Einheiten in zwei benachbarte P-P Bindungen des cyclo-P$_5$ Ringes gebildet. Außerdem wurde die Reaktivität von [Cp*Fe(η^5-P$_5$)] hinsichtlich des analogen Di-Germylens [LGe-GeL] untersucht, was zur Bildung von [(LGe)$_2${(μ-η^4-P$_5$)FeCp*}] (18) führte. Im Gegensatz zu Komplex 17 bleibt der cyclo-P$_5$ Ring in Komplex 18 intakt. Komplex 18 wird durch die formale Oxidation der Ge-Atome von der Oxidationsstufe +1 zu +2 gebildet, was zu zwei [LGe]$^+$-Einheiten und einer zweifach reduzierten [(η^4-P$_5$)FeCp*]$^{2-}$-Einheit führt. NMR-spektroskopische Untersuchungen ergaben eine langsame Zersetzung von Komplex 18 bei Raumtemperatur. Eine Röntgenstrukturanalyse des entsprechenden Thermolyseproduktes ergab die Ausbildung des Komplexes [(LGe){(μ-η^3-P$_5$)(η^1-LGe)FeCp*}] (19), der sich bedingt durch eine 1,2-migratorische Umlagerung einer [LGe]$^+$-Einheit am zyklischen P$_5$-Ring gebildet hat. Die [LGe]$^+$-Einheit koordiniert anstelle eines P-Atoms an das Fe-Zentrum. Die Reaktion von

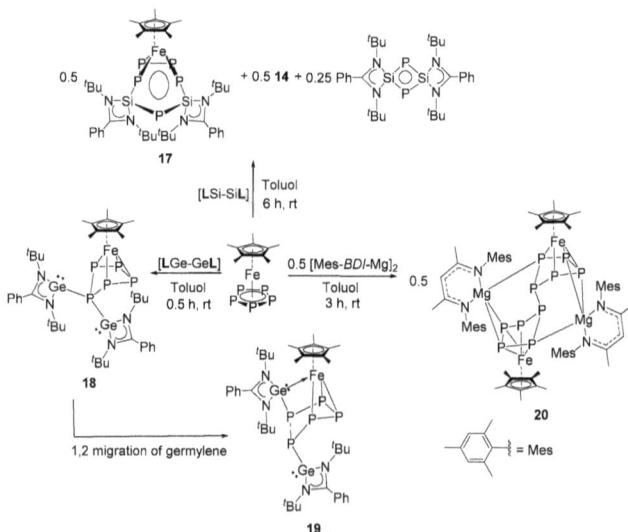

Schema 6.11: Reaktivität von [Cp*Fe(η^5-P$_5$)] mit niedervalenten Si-, Ge-, und Mg.

144

[Cp*Fe(η^5-P$_5$)] mit dem Mg(I)-Komplex [Mes-BDI-Mg^I]$_2$ (Mes-BDI = (2,4,6-Me$_3$C$_6$H$_3$NCMe)$_2$CH) führte zum Komplex [(Mes-BDI-Mg^{II})$_2$(μ-η^4-η^4-P$_{10}$)(FeCp*)$_2$] (20), welcher einen zentrale P$_{10}^{2-}$ Kern besitzt (Schema 6.11).

Weiterhin wurde die Reaktivität von [Cp*Fe(η^5-P$_5$)] mit einwertigen Aluminiumkomplexen untersucht. Die Reaktion von [Dipp-BDI-Al^I] (Dipp-BDI = (2,6-^{i}Pr$_2$C$_6$H$_3$NCMe)$_2$CH) und [Cp*Fe(η^5-P$_5$)] führte zur Bildung des Komplexes [(Dipp-BDI-Al^{III})(μ-η^3-η^4-P$_5$)FeCp*] (21) (Schema 6.12). Der briefumschlagförmige cyclo-P$_5$ Ring koordiniert hierbei η^3-an die [Dipp-BDI-Al^{II}]$^{2+}$-Einheit und η^4- an die [Cp*Fe]$^+$-Einheit, wodurch ein neutraler heterometallischer Al-Fe Tripledecker-Komplex entsteht. Die Reaktion des tetrameren, einwertigen [(Cp*Al^I)$_4$] mit [Cp*Fe(η^5-P$_5$)] bei 80 °C führt zum Aufbau des tetrametallischen Al-Fe Polyphosphid-Clusters [(P)(Cp*Al^{III})$_3$(μ-η^2-η^2-η^2-η^4-P$_4$)(FeCp*)] (22) (Schema 6.12). Der cyclo-P$_5$ Ring wird durch die Insertion einer [Cp*Al^{III}]$^{2+}$-Einheiten in eine azyklische P$_4$-Kette und eine P-Einheit gespalten. Das azyklische P$_4$ koordiniert η^4- an die [Cp*Fe]$^+$-Einheit und η^2 an jede [Cp*Al^{III}]$^{2+}$-Einheit, das „freie" Phosphoratom wird durch die drei [Cp*Al^{III}]$^{2+}$-Einheiten koordiniert. Unter drastischeren Reaktionsbedingungen kann der cyclo-P$_5$ Ring, unter Ausbildung des Al-Fe Polyphosphides 23 und Al-Polyphosphides 24, vollständig abgebaut werden.

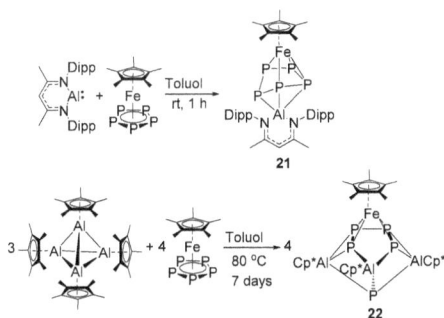

Schema 6.12: Synthese von Aluminium-Polyphosphiden.

7. References

[1]   F. A. Cotton, G. Wilkinson, *Advanced inorganic chemistry : a comprehensive text*, 4th ed., Wiley, New York; Chichester, **1980**.
[2]   G. B. Kauffman, *J. Chem. Educ.* **1979**, *56*, 496-499.
[3]   A. Werner, *Neuere Anschauungen auf dem Gebiete der anorganischen Chemie*, F. Vieweg und Sohn, Braunschweig, **1913**.
[4]   G. B. Kauffman, *Acs Sym Ser* **1994**, *565*, 3-33.
[5]   R. H. Crabtree, *The organometallic chemistry of the transition metals*, 5th ed., Wiley-Blackwell, Oxford, **2009**.
[6]   B. D. Gupta, *Basic Organometallic Chemistry: Concepts, Syntheses and Applications*, Universities Press, **2011**.
[7]   a) T. J. Kealy, P. L. Pauson, *Nature* **1951**, *168*, 1039-1040; b) G. Wilkinson, M. Rosenblum, M. C. Whiting, R. B. Woodward, *J. Am. Chem. Soc.* **1952**, *74*, 2125-2126.
[8]   E. O. Fischer, A. Maasböl, *Angew. Chem. Int. Ed.* **1964**, *3*, 580-581.
[9]   P. Buchwalter, J. Rosé, P. Braunstein, *Chem. Rev.* **2015**, *115*, 28-126.
[10]   F. A. Cotton, R. A. Walton, *Multiple bonds between metal atoms*, 2nd ed., Clarendon Press ; New York : Oxford University Press, Oxford, **1993**.
[11]   J. C. Green, M. L. H. Green, G. Parkin, *Chem. Commun.* **2012**, *48*, 11481-11503.
[12]   L. F. Dahl, E. Ishishi, R. E. Rundle, *J. Chem. Phys.* **1957**, *26*, 1750-1751.
[13]   a) F. A. Cotton, N. F. Curtis, C. B. Harris, B. F. G. Johnson, S. J. Lippard, J. T. Mague, W. R. Robinson, J. S. Wood, *Science* **1964**, *145*, 1305-1307; b) F. A. Cotton, C. B. Harris, *Inorg. Chem.* **1965**, *4*, 330-333.
[14]   T. Nguyen, A. D. Sutton, M. Brynda, J. C. Fettinger, G. J. Long, P. P. Power, *Science* **2005**, *310*, 844-847.
[15]   P. Kalck, *Top Organometal Chem* **2016**, *59*, V-Xii.
[16]   a) J. A. Mata, F. E. Hahn, E. Peris, *Chem. Sci.* **2014**, *5*, 1723-1732; b) B. Wurster, D. Grumelli, D. Hötger, R. Gutzler, K. Kern, *J. Am. Chem. Soc.* **2016**, *138*, 3623-3626.
[17]   a) U. Helmstedt, S. Lebedkin, T. Höcher, S. Blaurock, E. Hey-Hawkins, *Inorg. Chem.* **2008**, *47*, 5815-5820; b) J. Fernández-Gallardo, B. T. Elie, F. J. Sulzmaier, M. Sanaú, J. W. Ramos, M. Contel, *Organometallics* **2014**, *33*, 6669-6681; c) S. Bestgen, C. Schoo, C. Zovko, R. Köppe, R. P. Kelly, S. Lebedkin, M. M. Kappes, P. W. Roesky, *Chem. Eur. J.* **2016**, *22*, 7115-7126.
[18]   B. G. Cooper, J. W. Napoline, C. M. Thomas, *Catal Rev.* **2012**, *54*, 1-40.
[19]   L. H. Gade, *Angew. Chem. Int. Ed.* **2000**, *39*, 2658-2678.
[20]   K. M. Gramigna, D. A. Dickie, B. M. Foxman, C. M. Thomas, *ACS Catal.* **2019**, *9*, 3153-3164.
[21]   J. P. Krogman, B. M. Foxman, C. M. Thomas, *J. Am. Chem. Soc.* **2011**, *133*, 14582-14585.
[22]   J. P. Krogman, M. W. Bezpalko, B. M. Foxman, C. M. Thomas, *Dalton Trans.* **2016**, *45*, 11182-11190.
[23]   B. Wu, K. M. Gramigna, M. W. Bezpalko, B. M. Foxman, C. M. Thomas, *Inorg. Chem.* **2015**, *54*, 10909-10917.
[24]   a) F. Völcker, F. M. Mück, K. D. Vogiatzis, K. Fink, P. W. Roesky, *Chem. Commun.* **2015**, *51*, 11761-11764; b) F. Völcker, P. W. Roesky, *Dalton Trans.* **2016**, *45*, 9429-9435.
[25]   S. D. Cotton, *Lanthanide and actinide chemistry*, John Wiley, Chichester, **2006**.
[26]   M. N. Bochkarev, *Coord. Chem. Rev.* **2004**, *248*, 835-851.
[27]   F. Nief, *Dalton Trans.* **2010**, *39*, 6589-6598.
[28]   W. J. Evans, *Organometallics* **2016**, *35*, 3088-3100.
[29]   a) A. Krief, A.-M. Laval, *Chem. Rev.* **1999**, *99*, 745-778; b) P. G. Steel, *J. Chem. Soc., Perkin Trans. 1* **2001**, 2727-2751.

[30]   E. O. Fischer, H. Fischer, *Angew. Chem. Int. Ed.* **1964**, *3*, 132-133.
[31]   G. W. Watt, E. W. Gillow, *J. Am. Chem. Soc.* **1969**, *91*, 775-776.
[32]   W. J. Evans, I. Bloom, W. E. Hunter, J. L. Atwood, *J. Am. Chem. Soc.* **1981**, *103*, 6507-6508.
[33]   W. J. Evans, J. W. Grate, H. W. Choi, I. Bloom, W. E. Hunter, J. L. Atwood, *J. Am. Chem. Soc.* **1985**, *107*, 941-946.
[34]   C. Jones, *Nat. Rev. Chem.* **2017**, *1*, 0059.
[35]   W. J. Evans, L. A. Hughes, T. P. Hanusa, *J. Am. Chem. Soc.* **1984**, *106*, 4270-4272.
[36]   W. J. Evans, T. A. Ulibarri, J. W. Ziller, *J. Am. Chem. Soc.* **1988**, *110*, 6877-6879.
[37]   S. N. Konchenko, N. A. Pushkarevsky, M. T. Gamer, R. Köppe, H. Schnöckel, P. W. Roesky, *J. Am. Chem. Soc.* **2009**, *131*, 5740-5741.
[38]   C. Schoo, S. Bestgen, A. Egeberg, J. Seibert, S. N. Konchenko, C. Feldmann, P. W. Roesky, *Angew. Chem. Int. Ed.* **2019**, *58*, 4386-4389.
[39]   C. Schoo, S. Bestgen, A. Egeberg, S. Klementyeva, C. Feldmann, S. N. Konchenko, P. W. Roesky, *Angew. Chem. Int. Ed.* **2018**, *57*, 5912-5916.
[40]   W. J. Evans, C. A. Seibel, J. W. Ziller, *Inorg. Chem.* **1998**, *37*, 770-776.
[41]   W. J. Evans, J. W. Grate, L. A. Hughes, H. Zhang, J. L. Atwood, *J. Am. Chem. Soc.* **1985**, *107*, 3728-3730.
[42]   S. V. Klementyeva, N. Arleth, K. Meyer, S. N. Konchenko, P. W. Roesky, *New J. Chem.* **2015**, *39*, 7589-7594.
[43]   W. J. Evans, S. L. Gonzales, J. W. Ziller, *J. Am. Chem. Soc.* **1991**, *113*, 9880-9882.
[44]   W. J. Evans, G. W. Rabe, J. W. Ziller, R. J. Doedens, *Inorg. Chem.* **1994**, *33*, 2719-2726.
[45]   W. J. Evans, T. A. Ulibarri, J. W. Ziller, *J. Am. Chem. Soc.* **1990**, *112*, 2314-2324.
[46]   a) C. Schoo, S. Bestgen, R. Köppe, S. N. Konchenko, P. W. Roesky, *Chem. Commun.* **2018**, *54*, 4770-4773; b) D. Werner, X. Zhao, S. P. Best, L. Maron, P. C. Junk, G. B. Deacon, *Chem. Eur. J.* **2017**, *23*, 2084-2102.
[47]   F. T. Edelmann, D. M. M. Freckmann, H. Schumann, *Chem. Rev.* **2002**, *102*, 1851-1896.
[48]   M. L. Cole, P. C. Junk, *Chem. Commun.* **2005**, 2695-2697.
[49]   M. L. Cole, G. B. Deacon, C. M. Forsyth, P. C. Junk, K. Konstas, J. Wang, H. Bittig, D. Werner, *Chem. Eur. J.* **2013**, *19*, 1410-1420.
[50]   Y.-Z. Ma, S. Bestgen, M. T. Gamer, S. N. Konchenko, P. W. Roesky, *Angew. Chem. Int. Ed.* **2017**, *56*, 13249-13252.
[51]   J. Emsley, *The elements*, 2nd ed., Clarendon, **1991**.
[52]   C. E. Plečnik, S. Liu, S. G. Shore, *Acc. Chem. Res.* **2003**, *36*, 499-508.
[53]   R. Green, A. C. Walker, M. P. Blake, P. Mountford, *Polyhedron* **2016**, *116*, 64-75.
[54]   a) P. L. Arnold, J. McMaster, S. T. Liddle, *Chem. Commun.* **2009**, 818-820; b) T. Pugh, N. F. Chilton, R. A. Layfield, *Angew. Chem. Int. Ed.* **2016**, *55*, 11082-11085; c) C. P. Burns, X. Yang, S. Sung, J. D. Wofford, N. S. Bhuvanesh, M. B. Hall, M. Nippe, *Chem. Commun.* **2018**, *54*, 10893-10896.
[55]   a) T. D. Tilley, R. A. Andersen, *J. Chem. Soc., Chem. Commun.* **1981**, 985-986; b) T. D. Tilley, R. Andersen, *J. Am. Chem. Soc.* **1982**, *104*, 1772-1774; c) J. M. Boncella, R. A. Andersen, *Inorg. Chem.* **1984**, *23*, 432-437; d) A. Recknagel, A. Steiner, S. Brooker, D. Stalke, F. T. Edelmann, *Chem. Ber.* **1991**, *124*, 1373-1375; e) C. P. Burns, X. Yang, J. D. Wofford, N. S. Bhuvanesh, M. B. Hall, M. Nippe, *Angew. Chem. Int. Ed.* **2018**, *57*, 8144-8148.
[56]   a) C. E. Plečnik, S. Liu, X. Chen, E. A. Meyers, S. G. Shore, *J. Am. Chem. Soc.* **2004**, *126*, 204-213; b) M. P. Blake, N. Kaltsoyannis, P. Mountford, *J. Am. Chem. Soc.* **2011**, *133*, 15358-15361; c) *Chem. Commun.* **2013**, *49*, 3315-3317.
[57]   J. M. Boncella, R. A. Andersen, *J. Chem. Soc., Chem. Commun.* **1984**, 809-810.
[58]   G. B. Deacon, Z. Guo, P. C. Junk, J. Wang, *Angew. Chem. Int. Ed.* **2017**, *56*, 8486-8489.
[59]   A. C. Hillier, A. Sella, M. R. J. Elsegood, *J. Organomet. Chem.* **1999**, *588*, 200-204.

References

[60] S. N. Konchenko, T. Sanden, N. A. Pushkarevsky, R. Köppe, P. W. Roesky, *Chem. Eur. J.* **2010**, *16*, 14278-14280.

[61] H. Deng, S. G. Shore, *J. Am. Chem. Soc.* **1991**, *113*, 8538-8540.

[62] P. V. Poplaukhin, X. Chen, E. A. Meyers, S. G. Shore, *Inorg. Chem.* **2006**, *45*, 10115-10125.

[63] a) A. Togni, T. Hayashi, *Ferrocenes : homogeneous catalysis : organic synthesis: materials science*, VCH, Weinheim, Germany; Cambridge, **1995**; b) P. Nguyen, P. Gómez-Elipe, I. Manners, *Chem. Rev.* **1999**, *99*, 1515-1548; c) P. Štěpnička, *Ferrocenes: ligands, materials and biomolecules*, Wiley; Chichester: John Wiley, Hoboken, N.J., **2008**.

[64] a) W. P. Freeman, T. D. Tilley, A. L. Rheingold, R. L. Ostrander, *Angew. Chem. Int. Ed.* **1993**, *32*, 1744-1745; b) W. P. Freeman, T. D. Tilley, A. L. Rheingold, *J. Am. Chem. Soc.* **1994**, *116*, 8428-8429; c) V. Y. Lee, R. Kato, M. Ichinohe, A. Sekiguchi, *J. Am. Chem. Soc.* **2005**, *127*, 13142-13143; d) V. Y. Lee, R. Kato, A. Sekiguchi, A. Krapp, G. Frenking, *J. Am. Chem. Soc.* **2007**, *129*, 10340-10341; e) W. Wang, S. Yao, C. van Wüllen, M. Driess, *J. Am. Chem. Soc.* **2008**, *130*, 9640-9641; f) M. Saito, M. Sakaguchi, T. Tajima, K. Ishimura, S. Nagase, M. Hada, *Science* **2010**, *328*, 339-342; g) M. Saito, *Acc. Chem. Res.* **2018**, *51*, 160-169.

[65] H. V. Ly, J. Moilanen, H. M. Tuononen, M. Parvez, R. Roesler, *Chem. Commun.* **2011**, *47*, 8391-8393.

[66] W. Siebert, *Angew. Chem. Int. Ed.* **1985**, *24*, 943-958.

[67] G. De Lauzon, B. Deschamps, J. Fischer, F. Mathey, A. Mitschler, *J. Am. Chem. Soc.* **1980**, *102*, 994-1000.

[68] A. J. Ashe, T. R. Diephouse, J. W. Kampf, S. M. Al-Taweel, *Organometallics* **1991**, *10*, 2068-2071.

[69] A. J. Ashe, J. W. Kampf, S. M. Al-Taweel, *J. Am. Chem. Soc.* **1992**, *114*, 372-374.

[70] a) M. Baudler, *Angew. Chem. Int. Ed.* **1982**, *21*, 492-512; b) H. Grützmacher, *Z. Anorg. Allg. Chem.* **2012**, *638*, 1877-1879.

[71] O. J. Scherer, T. Brück, *Angew. Chem. Int. Ed.* **1987**, *26*, 59-59.

[72] M. Scheer, *Dalton Trans.* **2008**, 4372-4386.

[73] J. Bai, A. V. Virovets, M. Scheer, *Angew. Chem. Int. Ed.* **2002**, *41*, 1737-1740.

[74] J. Bai, A. V. Virovets, M. Scheer, *Science* **2003**, *300*, 781-783.

[75] W. I. F. David, R. M. Ibberson, J. C. Matthewman, K. Prassides, T. J. S. Dennis, J. P. Hare, H. W. Kroto, R. Taylor, D. R. M. Walton, *Nature* **1991**, *353*, 147-149.

[76] M. Scheer, A. Schindler, R. Merkle, B. P. Johnson, M. Linseis, R. Winter, C. E. Anson, A. V. Virovets, *J. Am. Chem. Soc.* **2007**, *129*, 13386-13387.

[77] C. Schwarzmaier, A. Schindler, C. Heindl, S. Scheuermayer, E. V. Peresypkina, A. V. Virovets, M. Neumeier, R. Gschwind, M. Scheer, *Angew. Chem. Int. Ed.* **2013**, *52*, 10896-10899.

[78] V. Peresypkina Eugenia, C. Heindl, A. Schindler, M. Bodensteiner, V. Virovets Alexander, M. Scheer, in *Z. Kristallogr. Cryst. Mater.*, Vol. 229, **2014**, p. 735.

[79] M. Scheer, L. J. Gregoriades, A. V. Virovets, W. Kunz, R. Neueder, I. Krossing, *Angew. Chem. Int. Ed.* **2006**, *45*, 5689-5693.

[80] R. F. Winter, W. E. Geiger, *Organometallics* **1999**, *18*, 1827-1833.

[81] M. V. Butovskiy, G. Balázs, M. Bodensteiner, E. V. Peresypkina, A. V. Virovets, J. Sutter, M. Scheer, *Angew. Chem. Int. Ed.* **2013**, *52*, 2972-2976.

[82] T. Li, J. Wiecko, N. A. Pushkarevsky, M. T. Gamer, R. Köppe, S. N. Konchenko, M. Scheer, P. W. Roesky, *Angew. Chem. Int. Ed.* **2011**, *50*, 9491-9495.

[83] T. Li, M. T. Gamer, M. Scheer, S. N. Konchenko, P. W. Roesky, *Chem. Commun.* **2013**, *49*, 2183-2185.

[84] C. Schoo, S. Bestgen, M. Schmidt, S. N. Konchenko, M. Scheer, P. W. Roesky, *Chem. Commun.* **2016**, *52*, 13217-13220.

[85] H. Krauss, G. Balázs, M. Bodensteiner, M. Scheer, *Chem. Sci.* **2010**, *1*, 337-342.

References

[86] E. Mädl, M. V. Butovskii, G. Balázs, E. V. Peresypkina, A. V. Virovets, M. Seidl, M. Scheer, *Angew. Chem. Int. Ed.* **2014**, *53*, 7643-7646.

[87] E. Mädl, E. Peresypkina, A. Y. Timoshkin, M. Scheer, *Chem. Commun.* **2016**, *52*, 12298-12301.

[88] D. Bourissou, O. Guerret, F. P. Gabbaï, G. Bertrand, *Chem. Rev.* **2000**, *100*, 39-92.

[89] H.-W. Wanzlick, H.-J. Schönherr, *Angew. Chem. Int. Ed.* **1968**, *7*, 141-142.

[90] A. Igau, H. Grutzmacher, A. Baceiredo, G. Bertrand, *J. Am. Chem. Soc.* **1988**, *110*, 6463-6466.

[91] A. J. Arduengo, R. L. Harlow, M. Kline, *J. Am. Chem. Soc.* **1991**, *113*, 361-363.

[92] J. Vignolle, X. Cattoën, D. Bourissou, *Chem. Rev.* **2009**, *109*, 3333-3384.

[93] H. V. Huynh, *Chem. Rev.* **2018**, *118*, 9457-9492.

[94] M. N. Hopkinson, C. Richter, M. Schedler, F. Glorius, *Nature* **2014**, *510*, 485.

[95] V. Nesterov, D. Reiter, P. Bag, P. Frisch, R. Holzner, A. Porzelt, S. Inoue, *Chem. Rev.* **2018**, *118*, 9678-9842.

[96] a) J. Cheng, L. Wang, P. Wang, L. Deng, *Chem. Rev.* **2018**, *118*, 9930-9987; b) A. A. Danopoulos, T. Simler, P. Braunstein, *Chem. Rev.* **2019**, *119*, 3730-3961.

[97] P. L. Arnold, I. J. Casely, *Chem. Rev.* **2009**, *109*, 3599-3611.

[98] a) F. E. Hahn, *Chem. Rev.* **2018**, *118*, 9455-9456; b) E. Peris, *Chem. Rev.* **2018**, *118*, 9988-10031; c) W. Wang, L. Cui, P. Sun, L. Shi, C. Yue, F. Li, *Chem. Rev.* **2018**, *118*, 9843-9929.

[99] A. Sekiguchi, T. Tanaka, M. Ichinohe, K. Akiyama, S. Tero-Kubota, *J. Am. Chem. Soc.* **2003**, *125*, 4962-4963.

[100] M. Weidenbruch, *Coord. Chem. Rev.* **1994**, *130*, 275-300.

[101] M. Denk, R. Lennon, R. Hayashi, R. West, A. V. Belyakov, H. P. Verne, A. Haaland, M. Wagner, N. Metzler, *J. Am. Chem. Soc.* **1994**, *116*, 2691-2692.

[102] a) S. Yao, Y. Xiong, M. Driess, *Organometallics* **2011**, *30*, 1748-1767; b) S. S. Sen, S. Khan, P. P. Samuel, H. W. Roesky, *Chem. Sci.* **2012**, *3*, 659-682; c) T. Chu, G. I. Nikonov, *Chem. Rev.* **2018**, *118*, 3608-3680.

[103] a) A. J. Downs, *Chemistry of aluminium, gallium, indium and thallium*, Blackie Academic & Professional, **1993**; b) S. Aldridge, A. J. Downs, *The chemistry of the group 13 metals Aluminium, Gallium, Indium, and Thallium : chemical patterns and peculiarities*, Wiley, Hoboken, N.J., **2011**.

[104] P. P. Power, *Nature* **2010**, *463*, 171.

[105] R. L. Melen, *Science* **2019**, *363*, 479-484.

[106] P. Bag, C. Weetman, S. Inoue, *Angew. Chem. Int. Ed.* **2018**, *57*, 14394-14413.

[107] C. Dohmeier, C. Robl, M. Tacke, H. Schnöckel, *Angew. Chem. Int. Ed.* **1991**, *30*, 564-565.

[108] M. Tacke, H. Schnoeckel, *Inorg. Chem.* **1989**, *28*, 2895-2896.

[109] C. Dohmeier, D. Loos, H. Schnöckel, *Angew. Chem. Int. Ed.* **1996**, *35*, 129-149.

[110] S. Schulz, H. W. Roesky, H. J. Koch, G. M. Sheldrick, D. Stalke, A. Kuhn, *Angew. Chem. Int. Ed.* **1993**, *32*, 1729-1731.

[111] M. T. Gamer, P. W. Roesky, S. N. Konchenko, P. Nava, R. Ahlrichs, *Angew. Chem.* **2006**, *118*, 4558-4561.

[112] S. G. Minasian, J. L. Krinsky, V. A. Williams, J. Arnold, *J. Am. Chem. Soc.* **2008**, *130*, 10086-10087.

[113] S. González-Gallardo, T. Bollermann, R. A. Fischer, R. Murugavel, *Chem. Rev.* **2012**, *112*, 3136-3170.

[114] A. Ecker, E. Weckert, H. Schnöckel, *Nature* **1997**, *387*, 379-381.

[115] P. Henke, T. Pankewitz, W. Klopper, F. Breher, H. Schnöckel, *Angew. Chem. Int. Ed.* **2009**, *48*, 8141-8145.

[116] H. Schnöckel, *Chem. Rev.* **2010**, *110*, 4125-4163.

[117] C. Cui, H. W. Roesky, H.-G. Schmidt, M. Noltemeyer, H. Hao, F. Cimpoesu, *Angew. Chem. Int. Ed.* **2000**, *39*, 4274-4276.

References

[118] a) H. W. Roesky, *Inorg. Chem.* **2004**, *43*, 7284-7293; b) H. W. Roesky, S. S. Kumar, *Chem. Commun.* **2005**, 4027-4038.

[119] a) C. Dohmeier, H. Schnöckel, C. Robl, U. Schneider, R. Ahlrichs, *Angew. Chem. Int. Ed.* **1994**, *33*, 199-200; b) Y. Peng, H. Fan, H. Zhu, H. W. Roesky, J. Magull, C. E. Hughes, *Angew. Chem. Int. Ed.* **2004**, *43*, 3443-3445.

[120] A. Hofmann, T. Tröster, T. Kupfer, H. Braunschweig, *Chem. Sci.* **2019**, *10*, 3421-3428.

[121] a) F. Mohr, M. C. Jennings, R. J. Puddephatt, *Angew. Chem. Int. Ed.* **2004**, *43*, 969-971; b) S. G. Dunning, G. Nandra, A. D. Conn, W. Chai, R. E. Sikma, J. S. Lee, P. Kunal, J. E. Reynolds III, J.-S. Chang, A. Steiner, G. Henkelman, S. M. Humphrey, *Angew. Chem. Int. Ed.* **2018**, *57*, 9295-9299; c) N. D. Knöfel, C. Schweigert, T. J. Feuerstein, C. Schoo, N. Reinfandt, A.-N. Unterreiner, P. W. Roesky, *Inorg. Chem.* **2018**, *57*, 9364-9375.

[122] G. Amenuvor, J. Darkwa, B. C. E. Makhubela, *Catal. Sci. Technol.* **2018**, *8*, 2370-2380.

[123] C. Jobbágy, P. Baranyai, Á. Gömöry, A. Deák, *CrystEngComm* **2018**, *20*, 5935-5939.

[124] V. W.-W. Yam, E. C.-C. Cheng, *Chem. Soc. Rev.* **2008**, *37*, 1806-1813.

[125] K.-H. Thiele, H. Rau, *Z. Anorg. Allg. Chem.* **1967**, *353*, 127-134.

[126] a) M. Amin Hasan, N. Kumari, P. Kumar, B. Pathak, L. Mishra, *Polyhedron* **2013**, *50*, 306-313; b) B. Murugesapandian, P. W. Roesky, *Inorg. Chem.* **2011**, *50*, 1698-1704; c) M. T. Ng, T. C. Deivaraj, W. T. Klooster, G. J. McIntyre, J. J. Vittal, *Chem. Eur. J.* **2004**, *10*, 5853-5859.

[127] a) J. He, Y. Wang, W. Bi, X. Zhu, R. Cao, *J. Mol. Struct.* **2006**, *787*, 63-68; b) Y.-F. Zhou, R.-H. Wang, B.-L. Wu, R. Cao, M.-C. Hong, *J. Mol. Struct.* **2004**, *697*, 73-79.

[128] S. Ahrland, K. Dreisch, B. Norén, Å. Oskarsson, *Mater. Chem. Phys.* **1993**, *35*, 281-289.

[129] A. Rit, T. P. Spaniol, L. Maron, J. Okuda, *Angew. Chem. Int. Ed.* **2013**, *52*, 4664-4667.

[130] a) A. S. K. Hashmi, I. Braun, M. Rudolph, F. Rominger, *Organometallics* **2012**, *31*, 644-661; b) C. Sarcher, A. Lühl, F. C. Falk, S. Lebedkin, M. Kühn, C. Wang, J. Paradies, M. M. Kappes, W. Klopper, P. W. Roesky, *Eur. J. Inorg. Chem.* **2012**, *2012*, 5033-5042.

[131] a) P. Pyykkö, *Angew. Chem. Int. Ed.* **2004**, *43*, 4412-4456; b) H. Schmidbaur, A. Schier, *Chem. Soc. Rev.* **2008**, *37*, 1931-1951; c) A. S. K. Hashmi, I. Braun, P. Nösel, J. Schädlich, M. Wieteck, M. Rudolph, F. Rominger, *Angew. Chem. Int. Ed.* **2012**, *51*, 4456-4460; d) A. S. K. Hashmi, M. Wieteck, I. Braun, P. Nösel, L. Jongbloed, M. Rudolph, F. Rominger, *Adv. Synth. Catal.* **2012**, *354*, 555-562; e) H. Schmidbaur, A. Schier, *Chem. Soc. Rev.* **2012**, *41*, 370-412.

[132] M. Andrejić, R. A. Mata, *Phys. Chem. Chem. Phys.* **2013**, *15*, 18115-18122.

[133] J. Gil-Rubio, J. Vicente, *Chem. Eur. J.* **2018**, *24*, 32-46.

[134] a) T. D. Pasatoiu, C. Tiseanu, A. M. Madalan, B. Jurca, C. Duhayon, J. P. Sutter, M. Andruh, *Inorg. Chem.* **2011**, *50*, 5879-5889; b) S. V. Eliseeva, J. C. Bunzli, *Chem. Soc. Rev.* **2010**, *39*, 189-227.

[135] a) J. Wu, L. Zhao, L. Zhang, X.-L. Li, M. Guo, A. K. Powell, J. Tang, *Angew. Chem. Int. Ed.* **2016**, *55*, 15574-15578; b) J.-L. Liu, J.-Y. Wu, Y.-C. Chen, V. Mereacre, A. K. Powell, L. Ungur, L. F. Chibotaru, X.-M. Chen, M.-L. Tong, *Angew. Chem. Int. Ed.* **2014**, *53*, 12966-12970; c) A. Bhunia, M. T. Gamer, L. Ungur, L. F. Chibotaru, A. K. Powell, Y. Lan, P. W. Roesky, F. Menges, C. Riehn, G. Niedner-Schatteburg, *Inorg. Chem.* **2012**, *51*, 9589-9597.

[136] a) M. Yadav, A. Bhunia, S. K. Jana, P. W. Roesky, *Inorg. Chem.* **2016**, *55*, 2701-2708; b) A. Bhunia, M. A. Gotthardt, M. Yadav, M. T. Gamer, A. Eichhöfer, W. Kleist, P. W. Roesky, *Chem. Eur. J.* **2013**, *19*, 1986-1995.

[137] T. Pugh, N. F. Chilton, R. A. Layfield, *Angew. Chem. Int. Ed.* **2016**, *55*, 11082-11085.

[138] M. L. Cole, P. C. Junk, *Chem. Commun.* **2005**, 2695-2697.

[139] F. Y.-K. Lo, G. Longoni, P. Chini, L. D. Lower, L. F. Dahl, *J. Am. Chem. Soc.* **1980**, *102*, 7691-7701.

[140] R. Shannon, *Acta Crystallogr., Sect. A* **1976**, *32*, 751-767.

[141] T. D. Tilley, R. Andersen, *J. Am. Chem. Soc.* **1982**, *104*, 1772-1774.

[142] T. D. Tilley, R. A. Andersen, *J. Chem. Soc., Chem. Commun.* **1981**, 985-986.

References

[143] M. P. Blake, N. Kaltsoyannis, P. Mountford, *J. Am. Chem. Soc.* **2015**, *137*, 12352-12368.
[144] W. J. Evans, I. Bloom, J. W. Grate, L. A. Hughes, W. E. Hunter, J. L. Atwood, *Inorg. Chem.* **1985**, *24*, 4620-4623.
[145] G. B. Deacon, Z. Guo, P. C. Junk, J. Wang, *Angew. Chem. Int. Ed.* **2017**, *56*, 8486-8489.
[146] E. E. Benson, C. P. Kubiak, *Chem. Commun.* **2012**, *48*, 7374-7376.
[147] M. R. Churchill, K. N. Amoh, H. J. Wasserman, *Inorg. Chem.* **1981**, *20*, 1609-1611.
[148] A. Alvanipour, H. Zhang, J. L. Atwood, *J. Organomet. Chem.* **1988**, *358*, 295-300.
[149] R. Yadav, T. Simler, M. T. Gamer, R. Köppe, P. W. Roesky, *Chem. Commun.* **2019**, *55*, 5765-5768.
[150] J. A. Baus, J. Poater, F. M. Bickelhaupt, R. Tacke, *Eur. J. Inorg. Chem.* **2017**, *2017*, 186-191.
[151] C. E. Kefalidis, S. Essafi, L. Perrin, L. Maron, *Inorg. Chem.* **2014**, *53*, 3427-3433.
[152] L. L. Padolik, J. C. Gallucci, A. Wojcicki, *J. Am. Chem. Soc.* **1993**, *115*, 9986-9996.
[153] M. J. Bennett, W. A. G. Graham, R. A. Smith, R. P. Stewart, *J. Am. Chem. Soc.* **1973**, *95*, 1684-1686.
[154] B. Blank, M. Colling-Hendelkens, C. Kollann, K. Radacki, D. Rais, K. Uttinger, G. R. Whittell, H. Braunschweig, *Chem. Eur. J.* **2007**, *13*, 4770-4781.
[155] J. M. Boncella, R. A. Andersen, *Inorg. Chem.* **1984**, *23*, 432-437.
[156] a) R. Ahlrichs, M. Bär, M. Häser, H. Horn, C. Kölmel, *Chem. Phys. Lett.* **1989**, *162*, 165-169; b) J. P. Perdew, *Phys. Rev. B* **1986**, *33*, 8822-8824; c) A. D. Becke, *Phys. Rev. A* **1988**, *38*, 3098-3100; d) J. P. Perdew, *Phys. Rev. B* **1986**, *34*, 7406-7406; e) M. Sierka, A. Hogekamp, R. Ahlrichs, *J. Chem. Phys.* **2003**, *118*, 9136-9148.
[157] a) F. Weigend, M. Häser, H. Patzelt, R. Ahlrichs, *Chem. Phys. Lett.* **1998**, *294*, 143-152; b) F. Weigend, R. Ahlrichs, *Phys. Chem. Chem. Phys.* **2005**, *7*, 3297-3305.
[158] R. Ahlrichs, C. Ehrhardt, *Chemie in unserer Zeit* **1985**, *19*, 120-124.
[159] T. Lu, F. Chen, *J. Comput. Chem.* **2012**, *33*, 580-592.
[160] R. Heinzmann, R. Ahlrichs, *Theor. Chem. Acc.* **1976**, *42*, 33-45.
[161] a) R. F. W. Bader, *Atoms in molecules : a quantum theory*, Clarendon Press, Oxford, **1990**; b) C. F. Matta, R. J. Gillespie, *J. Chem. Educ.* **2002**, *79*, 1141.
[162] G. Märkl, *Angew. Chem. Int. Ed.* **1965**, *4*, 1023-1038.
[163] L. Bialy, H. Waldmann, *Angew. Chem. Int. Ed.* **2005**, *44*, 3814-3839.
[164] J.-L. Montchamp, *Acc. Chem. Res.* **2014**, *47*, 77-87.
[165] W. Tang, X. Zhang, *Chem. Rev.* **2003**, *103*, 3029-3070.
[166] a) B. A. Arbuzov, G. N. Nikonov, O. A. Erastov, S. N. Ignatyeva, A. S. Balueva, A. A. Karasik, *Phosphorus Sulfur* **1990**, *49-50*, 271-274; b) H. H. Karsch, R. Richter, A. Schier, M. Heckel, R. Ficker, W. Hiller, *J. Organomet. Chem.* **1995**, *501*, 167-177; c) E. Herdtweck, F. Jäkle, M. Wagner, *Organometallics* **1997**, *16*, 4737-4745.
[167] M. Scheer, G. Balázs, A. Seitz, *Chem. Rev.* **2010**, *110*, 4236-4256.
[168] W. Lu, K. Xu, Y. Li, H. Hirao, R. Kinjo, *Angew. Chem. Int. Ed.* **2018**, *57*, 15691-15695.
[169] a) J. D. Masuda, W. W. Schoeller, B. Donnadieu, G. Bertrand, *J. Am. Chem. Soc.* **2007**, *129*, 14180-14181; b) C. L. Dorsey, B. M. Squires, T. W. Hudnall, *Angew. Chem. Int. Ed.* **2013**, *52*, 4462-4465.
[170] a) M. Driess, A. D. Fanta, D. R. Powell, R. West, *Angew. Chem. Int. Ed.* **1989**, *28*, 1038-1040; b) Y. Xiong, S. Yao, M. Brym, M. Driess, *Angew. Chem. Int. Ed.* **2007**, *46*, 4511-4513; c) S. Khan, R. Michel, S. S. Sen, H. W. Roesky, D. Stalke, *Angew. Chem. Int. Ed.* **2011**, *50*, 11786-11789; d) S. S. Sen, S. Khan, H. W. Roesky, D. Kratzert, K. Meindl, J. Henn, D. Stalke, J.-P. Demers, A. Lange, *Angew. Chem. Int. Ed.* **2011**, *50*, 2322-2325.
[171] a) S. Welsch, L. J. Gregoriades, M. Sierka, M. Zabel, A. V. Virovets, M. Scheer, *Angew. Chem. Int. Ed.* **2007**, *46*, 9323-9326; b) M. Scheer, L. J. Gregoriades, R. Merkle, B. P. Johnson, F. Dielmann, *Phosphorus Sulfur* **2008**, *183*, 504-508; c) M. Scheer, A. Schindler, C. Gröger, A. V. Virovets, E. V. Peresypkina, *Angew. Chem. Int. Ed.* **2009**, *48*, 5046-5049; d) M. Scheer, A. Schindler, J. Bai, B. P. Johnson, R. Merkle, R. Winter, A. V. Virovets, E. V. Peresypkina, V. A. Blatov, M. Sierka, H. Eckert,

References

Chem. Eur. J. **2010**, *16*, 2092-2107; e) S. Welsch, C. Gröger, M. Sierka, M. Scheer, *Angew. Chem. Int. Ed.* **2011**, *50*, 1435-1438; f) F. Dielmann, A. Schindler, S. Scheuermayer, J. Bai, R. Merkle, M. Zabel, A. V. Virovets, E. V. Peresypkina, G. Brunklaus, H. Eckert, M. Scheer, *Chem. Eur. J.* **2012**, *18*, 1168-1179; g) M. Fleischmann, S. Welsch, H. Krauss, M. Schmidt, M. Bodensteiner, E. V. Peresypkina, M. Sierka, C. Gröger, M. Scheer, *Chem. Eur. J.* **2014**, *20*, 3759-3768.

[172]   X. Bantreil, S. P. Nolan, *Nat. Protoc.* **2010**, *6*, 69.

[173]   A. J. Arduengo, H. V. R. Dias, R. L. Harlow, M. Kline, *J. Am. Chem. Soc.* **1992**, *114*, 5530-5534.

[174]   P. Pyykkö, M. Atsumi, *Chem. Eur. J.* **2009**, *15*, 12770-12779.

[175]   S. Khan, S. S. Sen, H. W. Roesky, *Chem. Commun.* **2012**, *48*, 2169-2179.

[176]   S. S. Sen, S. Khan, H. W. Roesky, D. Kratzert, K. Meindl, J. Henn, D. Stalke, J.-P. Demers, A. Lange, *Angew. Chem. Int. Ed.* **2011**, *123*, 2370-2373.

[177]   a) C.-W. So, H. W. Roesky, J. Magull, R. B. Oswald, *Angew. Chem. Int. Ed.* **2006**, *45*, 3948-3950; b) S. S. Sen, H. W. Roesky, D. Stern, J. Henn, D. Stalke, *J. Am. Chem. Soc.* **2010**, *132*, 1123-1126.

[178]   a) M. Driess, R. Janoschek, *J. Mol. Struct.* **1994**, *313*, 129-139; b) S. Inoue, W. Wang, C. Präsang, M. Asay, E. Irran, M. Driess, *J. Am. Chem. Soc.* **2011**, *133*, 2868-2871; c) A. E. Seitz, M. Eckhardt, A. Erlebach, E. V. Peresypkina, M. Sierka, M. Scheer, *J. Am. Chem. Soc.* **2016**, *138*, 10433-10436.

[179]   a) S. Gómez-Ruiz, E. Hey-Hawkins, *Dalton Trans.* **2007**, 5678-5683; b) M. Donath, F. Hennersdorf, J. J. Weigand, *Chem. Soc. Rev.* **2016**, *45*, 1145-1172; c) A. Wiesner, S. Steinhauer, H. Beckers, C. Müller, S. Riedel, *Chem. Sci.* **2018**, *9*, 7169-7173; d) K. Schwedtmann, J. Haberstroh, S. Roediger, A. Bauzá, A. Frontera, F. Hennersdorf, J. J. Weigand, *Chem. Sci.* **2019**; e) C. G. P. Ziegler, T. M. Maier, S. Pelties, C. Taube, F. Hennersdorf, A. W. Ehlers, J. J. Weigand, R. Wolf, *Chem. Sci.* **2019**, *10*, 1302-1308.

[180]   S. S. Sen, J. Hey, R. Herbst-Irmer, H. W. Roesky, D. Stalke, *J. Am. Chem. Soc.* **2011**, *133*, 12311-12316.

[181]   S. Nagendran, S. S. Sen, H. W. Roesky, D. Koley, H. Grubmüller, A. Pal, R. Herbst-Irmer, *Organometallics* **2008**, *27*, 5459-5463.

[182]   X. Chen, T. Simler, R. Yadav, M. T. Gamer, R. Köppe, P. W. Roesky, *Chem. Commun.* **2019**.

[183]   K. C. Thimer, S. M. I. Al-Rafia, M. J. Ferguson, R. McDonald, E. Rivard, *Chem. Commun.* **2009**, 7119-7121.

[184]   S. S. Sen, A. Jana, H. W. Roesky, C. Schulzke, *Angew. Chem. Int. Ed.* **2009**, *48*, 8536-8538.

[185]   Y. Xiong, S. Yao, E. Szilvási, E. Ballestero-Martínez, H. Grützmacher, M. Driess, *Angew. Chem. Int. Ed.* **2017**, *56*, 4333-4336.

[186]   G. E. Quintero, I. Paterson-Taylor, N. H. Rees, J. M. Goicoechea, *Dalton Trans.* **2016**, *45*, 1930-1936.

[187]   Y.-L. Shan, B.-X. Leong, H.-W. Xi, R. Ganguly, Y. Li, K. H. Lim, C.-W. So, *Dalton Trans.* **2017**, *46*, 3642-3648.

[188]   a) J. S. Figueroa, C. C. Cummins, *Angew. Chem. Int. Ed.* **2005**, *44*, 4592-4596; b) S. Mitzinger, J. Bandemehr, K. Reiter, J. Scott McIndoe, X. Xie, F. Weigend, J. F. Corrigan, S. Dehnen, *Chem. Commun.* **2018**, *54*, 1421-1424.

[189]   S. P. Green, C. Jones, A. Stasch, *Science* **2007**, *318*, 1754-1757.

[190]   J. Overgaard, C. Jones, A. Stasch, B. B. Iversen, *J. Am. Chem. Soc.* **2009**, *131*, 4208-4209.

[191]   a) S. J. Bonyhady, D. Collis, G. Frenking, N. Holzmann, C. Jones, A. Stasch, *Nat. Chem.* **2010**, *2*, 865; b) R. Lalrempuia, C. E. Kefalidis, S. J. Bonyhady, B. Schwarze, L. Maron, A. Stasch, C. Jones, *J. Am. Chem. Soc.* **2015**, *137*, 8944-8947; c) S. J. Bonyhady, N. Holzmann, G. Frenking, A. Stasch, C. Jones, *Angew. Chem. Int. Ed.* **2017**, *56*, 8527-8531; d) S. J. Bonyhady, D. Collis, N. Holzmann, A. J. Edwards, R. O. Piltz, G. Frenking, A. Stasch, C. Jones, *Nat. Commun.* **2018**, *9*, 3079.

[192]   a) O. T. Summerscales, F. G. N. Cloke, P. B. Hitchcock, J. C. Green, N. Hazari, *Science* **2006**, *311*, 829-831; b) K. Yuvaraj, I. Douair, A. Paparo, L. Maron, C. Jones, *J. Am. Chem. Soc.* **2019**, *141*, 8764-8768.

[193]   I. R. R. Peloso, A. Rodríguez, E. Carmona, K. Freitag, C. Jones, A. Stasch, A. J. Boutland, F. Lips,, in *Inorg. Synth., Vol. 37* (Ed.: P. P. Power), John Wiley & Sons, Inc., **2018**, pp. 33-45.

[194]   M. Arrowsmith, M. S. Hill, A. L. Johnson, G. Kociok-Köhn, M. F. Mahon, *Angew. Chem. Int. Ed.* **2015**, *54*, 7882-7885.

[195]   C. R. Groom, I. J. Bruno, M. P. Lightfoot, S. C. Ward, *Acta Crystallogr., Sect. B* **2016**, *72*, 171-179.

[196]   S. Nagendran, H. W. Roesky, *Organometallics* **2008**, *27*, 457-492.

[197]   N. Kuhn, T. Kratz, *Synthesis* **1993**, *1993*, 561-562.

[198]   M. Stender, R. J. Wright, B. E. Eichler, J. Prust, M. M. Olmstead, H. W. Roesky, P. P. Power, *J. Chem. Soc., Dalton Trans.* **2001**, 3465-3469.

[199]   P. L. Watson, T. H. Tulip, I. Williams, *Organometallics* **1990**, *9*, 1999-2009.

[200]   K. Hirano, S. Urban, C. Wang, F. Glorius, *Org. Lett.* **2009**, *11*, 1019-1022.

[201]   a) G. Sheldrick, *Acta Crystallogr., Sect. A* **2008**, *64*, 112-122; b) G. Sheldrick, *Acta Crystallogr., Sect. C* **2015**, *71*, 3-8.

[202]   O. V. Dolomanov, L. J. Bourhis, R. J. Gildea, J. A. K. Howard, H. Puschmann, *J. Appl. Cryst.* **2009**, *42*, 339-341.

8. Appendix

8.1 Directory of Abbreviations

Bipy	2,2'-bipyridine
ca.	approximately
Calcd	calculated
Cp	cyclopentadienyl
Cp*	pentamethylcyclopentadienyl
CV	cyclovoltammetry
Dipp-*BDI*	$(2,6\text{-}^{i}Pr_2C_6H_3NCMe)_2CH$
DippForm	*N,N'*-bis(2,6-diisopropylphenyl)formamidinate
DME	dimethoxyethane
DFT	density functional theory
Et_2O	diethylether
Et	ethyl
h	hour
Hz	Hertz
HOMO	highest occupied molecular orbital
^{i}Pr	isopropyl
IPr	$1,3\text{-bis}(2,6\text{-}^{i}Pr_2C_6H_3)$imidazol-2-ylidene
IR	infrared
ITMe	1,3,4,5-tetramethylimidazol-2-ylidene
K	Kelvin
L	$PhC(N^{t}Bu)_2$
Ln	lanthanides
LUMO	lowest unoccupied molecular orbital
Me	methyl
Mes-*BDI*	$(2,4,6\text{-}Me_3C_6H_3NCMe)_2CH$

mmol	millimole
NHC	N-heterocyclic carbene
NHSi	N-heterocyclic silylene
NMR	nuclear magnetic resonance
^{t}Bu	tertiary butyl
R	organic group
rt	room temperature
thf	tetrahydrofuran
tht	tetrahydrothiophene
TM	transition metals
XRD	X-ray diffraction

8.2 NMR Abbreviations

d	doublet
dd	doublet of doublet
J	coupling constant
m	multiplet
ppm	parts per million
qt	quintet
s	singlet
t	triplet
δ	chemical shift

8.3 IR Abbreviations

br	broad
m	medium
s	strong
vs	very strong
w	weak
vw	very weak

8.4 Directory of Compounds

1	$[(Bipy)Zn(p\text{-}O_2C(C_6H_4)PPh_2)_2]$
2a	$[(Bipy)Zn(p\text{-}O_2C(C_6H_4)PPh_2(AuCl))_2]$
2	$[(Bipy)_2Zn_3\{p\text{-}O_2C(C_6H_4)PPh_2(AuCl)\}_6]$
3	$[(Bipy)Zn(O_2C(C_2H_4)PPh_2)_2]$
4	$[(Bipy)_2Zn_3\{O_2C(C_2H_4)PPh_2(AuCl)\}_6]$
5	$[\{(DippForm)_2Sm^{III}\}_2\{(\mu_3\text{-}CO)_2(CO)_9Fe_3\}]$
6	$[\{(DippForm)_2Sm^{III}(thf)\}_2\{(\mu\text{-}CO)_2(CO)_2Co\}_2]$
7	$[\{(DippForm)_2Yb^{III}(thf)\}\{(\mu\text{-}CO)(CO)_3Co\}]$
8	$[\{(DippForm)_2Sm^{III}(thf)\}\{(\mu\text{-}CO)(CO)_4Mn\}]$
9	$[\{(DippForm)_2Yb^{III}(thf)\}\{(\mu\text{-}CO)(CO)_4Mn\}]$
10	$[\{(DippForm)_2Sm^{III}(thf)\}_2\{(\mu\text{-}\eta^2\text{-}CO)_2(\mu\text{-}\eta^1\text{-}CO)_2(CO)_4Re_2\}]$
11	$[\{(Cp^*)_2Sm^{III}\}_3\{(Cp^*)_2Sm^{III}(thf)\}\{(\mu\text{-}O_4C_4)(\mu\text{-}\eta^2\text{-}CO)_2(\mu\text{-}\eta^1\text{-}CO)(CO)_5Re_2\}]$
12	$[\{(Cp^*)_2Sm^{III}(thf)\}\{(\mu\text{-}CO)_2(CO)_3Mn\}]_n$
13	$[ITMe\{(\eta^4\text{-}P_5)FeCp^*\}]$
14	$[(\eta^4\text{-}P_4SiL)FeCp^*]$
15	$[LSi(Cl)=P\text{-}SiL(Cl)_2]$
16	$[\{LSi(N(SiMe_3)_2)\}\{(\eta^4\text{-}P_5)FeCp^*\}]$
17	$[\{(\eta^4\text{-}P_5(SiL)_2\}FeCp^*]$
18	$[(LGe)_2\{(\mu\text{-}\eta^4\text{-}P_5)FeCp^*\}]$
19	$[(LGe)\{(\mu\text{-}\eta^3\text{-}P_5)(\eta^1\text{-}GeL)FeCp^*\}]$
20	$[(Mes\text{-}BDI\text{-}Mg^{II})_2(\mu\text{-}\eta^4\text{-}\eta^4\text{-}P_{10})(FeCp^*)_2]$
21	$[Dipp\text{-}BDI\text{-}Al^{III}(\mu\text{-}\eta^4\text{-}P_5)FeCp^*]$
22	$[(P)(Cp^*Al^{III})_3(\mu\text{-}\eta^2\text{-}\eta^2\text{-}\eta^2\text{-}\eta^4\text{-}P_4)(FeCp^*)]$
23	$[(Cp^*Al^{III})_4(P_{10})(Cp^*Fe)_2]$
24	$[(Cp^*Al^{III})_6(P_6)]$

9. Curriculum Vitae

Name	Ravi Yadav
Date of Birth	14th January 1993
Place of Birth	Jhajjar, Haryana, India
Gender	Male
Marital Status	Single
Nationality	Indian

Education

Ph. D. in Chemistry	(July 2016 - present) Department of Chemistry, Karlsruhe Institute of Technology, Karlsruhe, Germany
Thesis Supervisor	Prof. Dr. Peter W. Roesky
Thesis Title	Synthesis of Heterometallic Zinc-Gold and Lanthanide-Transition Metal Carbonyl Complexes and Reactivity Study of Pentaphosphaferrocene Towards Low-Valent Main Group Species
Master of Science	(2016) Department of Chemistry, Indian Institute of Technology, Delhi, India
Bachelor of Science	(2014) Chemistry (Honors), Ramjas College, Delhi University, Delhi, India

Poster

Fischer-type Rhenacycle: CO Tetramerization Induced by a Divulent Lanthanide Complex in Rhenium Carbonyls

Ravi Yadav, Thomas Simler, Michael T. Gamer, Ralf Köppe, and Peter W. Roesky

The 13th International Conference on Heteroatom Chemistry (**ICHAC 2019**) in Prague from June 30 to July 5, 2019

Appendix

Presentation

Homogeneous Catalysts Based on the Rare-Earth Elements and related Systems.

Ravi Yadav (Representing Roesky Group)

IIT Indore- TU9 Second Workshop, From 5 to 6 October 2017 at IIT Indore, Simrol, Indore, India

Publications

1) R. Yadav, T. Simler, M. T. Gamer, R. Köppe, P. W. Roesky, *Chem. Commun.* **2019**, *55*, 5765-5768

2) X. Chen, T. Simler, R. Yadav, M. T. Gamer, R. Köppe, P. W. Roesky, *Chem. Commun.* **2019**, *55*, 9315-9318

3) T. Simler, T. J. Feuerstein, R. Yadav, M. T. Gamer, P. W. Roesky, *Chem. Commun.* **2019**, *55*, 222-225

4) X. Sun, T. Simler, R. Yadav, R. Köppe, P. W. Roesky, *J. Am. Chem. Soc.*, **2019**, *141*, 14987-14990

5) M. K. Sharma, D. Singh, P. Mahawar, R. Yadav, S. Nagendran, *Dalton Trans.* **2018**, *47*, 5943-5947

10. Acknowledgements

I would like to express a sincere thanks to my supervisor Prof. Dr. Peter W. Roesky for his excellent guidance, constant encouragement, optimism, and wonderful cooperation during every stage of my doctoral thesis. His expertise on the different areas of inorganic chemistry and very honest and critical review of my work helped me to understand of various aspects of synthetic organometallic chemistry. I am very grateful for getting an opportunity to work in his research group.

I am wholeheartedly thankful to Dr. Thomas Simler for his all around help and being an amazing colleague and a good friend. I am very fortunate to discuss and learn from him every day.

I am thankful to my good friend Bhupendra Goswami for his excellent help and it has been a great experience working with him. I am very thankful to my good friend Dr. Xiao Chen for great help and providing a wonderful working environment in the lab 429.

I would like to thank Dr. Ralf Köppe for his help with theoretical calculations.

I would like to thank Prof. Dr. Sergey Konchenko, Dr. Ying-Zhao Ma, and Dr. Vladimir Dodonov for demonstrating several techniques to handle air and moisture sensitive compounds.

I am very thankful to Sibylle Schneider for measuring crystals and for her very kind and helpful nature. I would like to thank Dr. Michael Gamer, Dr. Thomas Simler and Dr. Christoph Schoo for their help in solving the crystal structures.

I appreciate Mrs. Kayas and Mrs. Pendl for their assistance with paper and document work. I am thankful to Mrs. Berberich (NMR), Mrs. Smie (Mass), Mrs. Stößer (chemical ordering), Mrs. Klaassen (elemental analysis), Mrs. Leichle (chemical store), Mr. Munshi (glassblowing), Mr. Rieß, Mr. Lampert, and Mr. Kastner (mechanical workshop) for their help.

I would like to thank Dr. Thomas Simler, Bhupendra Goswami, Xiaofei Sun, and Xiao Chen for proofreading my thesis. I would like to thank Xiaofei Sun, Christina Zovko, Nicolai Knöfel, Sebastian Kaufmann, and Christoph Schoo for proofreading and German translation of my summary.

I am thankful to all of my collegues in AK Roesky for their help and providing a pleasant working environment.

I am thankful to SFB1176 for my doctoral fellowship.

I would like to thank my friends Akash, Raghu, Vaibhav, Ankit, Abhishek, Krishan, Deepak, Shrikesh, Rahul, Vishal, Rajan, Debotra, Tarachand, and Navid for their support.

Finally, I wish to pay tribute to my family who sacrificed their worldly interests to promote my education.